AF388891

Polyphenolic Antioxidants from Agri-Food Waste Biomass

Polyphenolic Antioxidants from Agri-Food Waste Biomass

Special Issue Editors

Dimitris P. Makris
Selin Şahin

MDPI • Basel • Beijing • Wuhan • Barcelona • Belgrade • Manchester • Tokyo • Cluj • Tianjin

Special Issue Editors

Dimitris P. Makris
University of Thessaly
Greece

Selin Şahin
Istanbul University-Cerrahpaşa
Turkey

Editorial Office
MDPI
St. Alban-Anlage 66
4052 Basel, Switzerland

This is a reprint of articles from the Special Issue published online in the open access journal *Antioxidants* (ISSN 2076-3921) (available at: https://www.mdpi.com/journal/antioxidants/special_issues/agri-food).

For citation purposes, cite each article independently as indicated on the article page online and as indicated below:

LastName, A.A.; LastName, B.B.; LastName, C.C. Article Title. *Journal Name* **Year**, *Article Number*, Page Range.

ISBN 978-3-03928-674-4 (Pbk)
ISBN 978-3-03928-675-1 (PDF)

Contents

About the Special Issue Editors

Dimitris P. Makris, Associate Professor, holds a B.S. in Œnology and Beverage Technology from the Technological Educational Institute (T.E.I.) of Athens (now University of West Attica), a M.S. in Œnology from the University of Burgundy, France), and a Ph.D. in Food Chemistry from Imperial College – University of London (U.K.). He has been employed as post-doctoral researcher in various institutes and, in 2010, he was appointed Lecturer of Food Biochemistry at the Department of Food Science and Nutrition, University of the Aegean. In 2014, he was appointed Assistant Professor in the same Department and in 2019 as Associate Professor at the Department of Food Science and Nutrition, University of Thessaly. He has participated in over than 10 research national and international programs, and he has over than 140 publications in international peer-reviewed journals and over than 6000 citations (Google Scholar). He is currently active Editorial Board member in several journals (Elsevier, MDPI, Hindawy, Omics). His research is mainly focused on valorization of food waste biomass, extraction technology, polyphenolic antioxidants, and natural pigments.

Selin Şahin, Dr., completed her Ph.D. at Istanbul University. She has an experience as a Guest Researcher with the United States Department of Agriculture (USDA). She is associate professor at the Department of Chemical Engineering. She has an academic experience in chemical engineering with special interests in novel separation methods, recovery of natural products, removal of pollutants in wastewater, purification, food additives, lipid oxidation, optimization, and statistics with 70 international scientific papers published in reputed journals with ove 1000 citations and participating to national/international congrees with 50 presentations

Editorial

Polyphenolic Antioxidants from Agri-Food Waste Biomass

Dimitris P. Makris [1,*] and **Selin Şahin** [2]

[1] Green Processes & Biorefinery Group, School of Agricultural Sciences, University of Thessaly, 43100 Karditsa, Greece
[2] Chemical Engineering Department, Istanbul University-Cerrahpaşa, 34320 Avcilar, Istanbul, Turkey; selins@istanbul.edu.tr
* Correspondence: dimitrismakris@uth.gr; Tel.: +30-244-106-4792

Received: 3 December 2019; Accepted: 5 December 2019; Published: 6 December 2019

As the world's population is rapidly expanding, environmental aggravation and bioresource depletion are becoming challenges of paramount importance. Agricultural production and the food supply chain are major sources of waste biomass, which poses an unprecedented risk to land and water pollution, and eventually, public health. However, agri-food processing residues, owed to their peculiar composition, are also recognized as materials of high biorefinery potency, offering a range of opportunities for sustainable production of food, feed, chemicals, and energy.

In such a framework, the valorization of by-products and wastes originating from the agricultural and food sector for the production of a broad spectrum of high value-added products has emerged as a key strategy. The exploitation of agri-food side streams by implementing eco-friendly and cost-effective technologies is now considered as a primary route towards zero-waste production, and the design and commercialization of novel bio-based formulations has proven to be significantly more profitable than biofuel production [1].

Polyphenols are bioactive secondary plant metabolites possessing versatile properties, including long-term protection from cardiovascular diseases, chemoprotective activity, antioxidant and anti-inflammatory potency. The recovery of polyphenolic substances from agri-food wastes is a primary target in the higher-value options of biorefining, and numerous investigations have been and are being carried out towards this direction [2]. The objective of this special issue is a compilation of the most recent, state-of-the-art studies pertaining to the valorization of food processing wastes to highlight cutting-edge advances in the field.

Kim et al. [3] investigate the phenanthrene content of waste peels from a popular Asian food, the Chinese yam (*Dioscorea batatas*). Phenanthrenes are major polyphenolic substances in yam, and their content may define its nutrition value. The authors, after performing a detailed High-performance liquid chromatography (HPLC) analysis, conclude that yam peels may be a source rich in phenanthrenes, and thus they may merit greater attention as a bioresource of functional phytochemicals.

Alañón et al. [4] study the extracts from mango by-products (peel, husk seed, and seed) for potential bioactivity. The results drawn showed that mango seed extract afforded a 72% percentage inhibition of platelet aggregation induced by adenosine 5'-diphosphate (ADP) agonist in a dose-dependent manner. The most potent extract contained monogalloyl compounds, tetra- and penta-galloylglucose, ellagic acid, mangiferin, and benzophenones such as maclurin derivatives and iriflophenone glucoside. Out of these compounds, mangiferin exhibited an inhibitory effect of 31%, suggesting its key role as one of the main contributors to the antiplatelet activity of mango seed. The authors suggest that mango seed could be a bioresource of compounds with antiplatelet properties and may be used for designing functional foods.

Di Mauro et al. [5] examine the possibility of using olive mill wastewater (OMW) to produce ophthalmic nutraceutical formulations. Various adsorbents were tested to selectively recover a polyphenol-rich fraction, which was then assayed for cytotoxicity and antioxidant/anti-inflammatory activities through in vitro tests. The results indicate that the fraction (0.01%) had no toxic effects and was able to protect cells against oxidant and inflammatory stimulus, reducing reactive oxygen species and TNF-α levels. The authors also prepared a novel stable ophthalmic hydrogel containing a polyphenolic fraction (0.01%) and assessed the technical and economic feasibility of the process at a pre-industrial level.

Birsan et al. [6] study the effective recovery of antioxidant polyphenols from light, dark, and mix brewer's spent grain (BSG) using conventional maceration, microwave, and ultrasound-assisted extraction. Irrespective of the extraction methods used, the saponification of BSG yielded higher polyphenols than in the crude extracts, with the EtOAc fractionations being the most effective. Microwave and ultrasound-assisted extractions did not improve the total polyphenol yield when compared to the conventional maceration method. The authors conclude that BSG light may be regarded as a sustainable, low-cost source of natural antioxidants, tapped for applications in the food and phytopharmaceutical industries.

Matos et al. [7] investigate winemaking waste streams as a material with cosmeceutical potential. The authors produced extracts from grape marc and wine lees, using solid–liquid (SL) extraction with and without microwave (MW) pretreatment, and assayed them for antioxidant activity through chemical (ORAC/HOSC/HORAC) and cell-based (keratinocytes—HaCaT; fibroblasts—HFF) tests. Their inhibitory capacity towards specific enzymes involved in skin ageing (elastase; MMP-1; tyrosinase) was also appraised. MW pretreatment prior to conventional SL extraction led to overall better outcomes. Red wine lees extracts presented the highest phenolic content and exhibited the highest antioxidant capacity, being also the most effective inhibitors of elastase, MMP-1, and tyrosinase. The authors argue that winemaking waste streams could be valuable sources of natural ingredients for cosmeceutical applications.

Martín-García et al. [8] focus their research on the establishment of ultrasound-assisted extraction of proanthocyanidin compounds from brewing spent grains using a sonotrode. Response surface methodology was used to study the effects of three factors, namely, solvent composition, time of extraction, and ultrasound power. The highest content of proanthocyanidins was obtained using 80/20 acetone/water (*v/v*), 55 min, and 400 W. The authors support that this methodology allows for the extraction of 1.01 mg/g dry weight of pronthocyanidins from brewer's spent grain, this value being more than two times higher than conventional extraction.

Abi-Khattar et al. [9] present an innovative technology for effective polyphenol recovery from olive leaves, namely Ired-Irrad®. In this study, optimization of infrared-assisted extraction was conducted using response surface methodology to intensify polyphenol recovery from olive leaves. The extraction efficiency using Ired-Irrad®, a newly patented infrared apparatus (IR), was compared to the water bath (WB) conventional extraction. Under optimal conditions, the total phenolic content yield was enhanced by more than 30% using IR as contrasted to WB, which required 27% more ethanol consumption. The extraction of two major phenolic compounds of the leaves, oleuropein and hydroxytyrosol, was intensified by 18% and 21%, respectively. IR extracts increased the antiradical activity by 25% and the antioxidant capacity by 51% compared to WB extracts. On the other hand, extracts of olive leaves obtained by both techniques exhibited equal effects regarding the inhibition of 20 strains of *Staphylococcus aureus*. Similarly, both extracts inhibited aflatoxin B1 (AFB1) secretion by *Aspergillus flavus*, with no growth inhibition of the fungus. The authors claim that this innovative technique allows for significantly reduced energy and solvent consumption while maintaining a similar quantity and quality of phenolic compounds as what is optimally obtained using WB.

Fernandez et al. [10] propose apple pomace as a sustainable food ingredient. The authors used acidified hot water extraction as a clean, feasible, and easy approach for the recovery of polyphenols. This technique allowed them to obtain 296 g of extract per kg of dry apple pomace, including 3.3 g of polyphenols and 281 g of carbohydrates. Ultrafiltration and solid-phase extraction using C18 cartridges of the hot water extract suggested that, in addition to the apple native polyphenols, polyphenols could also be present as complexes with carbohydrates. For the water-soluble polyphenols, antioxidant and anti-inflammatory effects were observed by inhibiting chemically generated hydroxyl radicals (OH•) and nitrogen monoxide radicals (NO•) produced in lipopolysaccharide-stimulated macrophages. The water-soluble polyphenols, when incorporated into yogurt formulations, were not affected by fermentation and improved the antioxidant properties of the final product.

Li et al. [11] propose the use of edible oils as effective means of recovering volatile and non-volatile compounds from rosemary since these materials can serve as food-grade solvents. Soybean oil could obtain the highest total phenolic compounds among 12 refined oils including grapeseed, rapeseed, peanut, sunflower, olive, avocado, almond, apricot, corn, wheat germ, and hazelnut oils. The addition of oil derivatives to soybean oils, such as glyceryl monooleate, glyceryl monostearate, diglycerides, and soy lecithin, not only significantly enhanced the oleo-extraction of non-volatile antioxidants by 66.7% approximately but also help to remarkably improve the solvation of volatile aroma compounds by 16% in refined soybean oils. These results were in good consistency with their relative solubilities predicted by the more sophisticated COSMO-RS (COnductor like Screening MOdel for Real Solvents) simulation. According to the authors, the use of vegetable oils and their derivatives as bio-based solvents for the extraction yield of natural antioxidants and flavors from rosemary show an important potential in up-scaling. Along with the integration of green techniques (e.g., ultrasound, microwave), such technologies can contribute towards zero-waste biorefinery from biomass waste and the production of high value-added extracts in future functional food and cosmetic applications.

Finally, Lakka et al. [12] deliver an investigation on saffron processing wastes as a bioresource of high value-added compounds and the development of a green extraction process for polyphenol recovery using a natural deep eutectic solvent. The study included an appraisal of the molar ratio of hydrogen bond donor/hydrogen bond acceptor in order to come up with the most efficient DES composed of L-lactic acid/glycine (5:1), and then optimization of the extraction process using response surface methodology. Under optimal conditions, the extraction yield in total polyphenols achieved was 132.43 ± 10.63 mg gallic acid equivalents per g of dry mass. The temperature assay suggested that extracts displayed maximum yield and antioxidant activity at 50–60 °C. Liquid chromatography-mass spectrometry analysis of the SPW extract obtained under optimal conditions showed that the predominant flavonol was kaempferol 3-*O*-sophoroside and the major anthocyanin delphinidin 3,5-di-*O*-glucoside. The results indicate that SPW extraction with the DES used is a green and efficient methodology and may afford extracts rich flavonols and anthocyanins, which are considered to be powerful antioxidants.

Conflicts of Interest: The authors declare no conflict of interest.

References

1. Otles, S.; Despoudi, S.; Bucatariu, C.; Kartal, C. Food waste management, valorization, and sustainability in the food industry. In *Food Waste Recovery*; Elsevier: London, UK, 2015; pp. 3–23.
2. Alexandre, E.M.; Castro, L.M.; Moreira, S.A.; Pintado, M.; Saraiva, J.A. Comparison of emerging technologies to extract high-added value compounds from fruit residues: Pressure-and electro-based technologies. *Food Eng. Rev.* **2017**, *9*, 190–212. [CrossRef]
3. Kim, M.; Gu, M.; Lee, J.; Chin, J.; Bae, J.; Hahn, D. Quantitative analysis of bioactive phenanthrenes in *Dioscorea batatas* decne peel, a discarded biomass from postharvest processing. *Antioxidants* **2019**, *8*, 541. [CrossRef] [PubMed]
4. Alañón, M.; Palomo, I.; Rodríguez, L.; Fuentes, E.; Arráez-Román, D.; Segura-Carretero, A. Antiplatelet activity of natural bioactive extracts from mango (*Mangifera indica* L.) and its by-products. *Antioxidants* **2019**, *8*, 517. [CrossRef] [PubMed]
5. Di Mauro, M.; Fava, G.; Spampinato, M.; Aleo, D.; Melilli, B.; Saita, M.; Centonze, G.; Maggiore, R.; D'Antona, N. Polyphenolic fraction from olive mill wastewater: Scale-up and in vitro studies for ophthalmic nutraceutical applications. *Antioxidants* **2019**, *8*, 462. [CrossRef] [PubMed]

6. Birsan, R.; Wilde, P.; Waldron, K.; Rai, D. Recovery of polyphenols from brewer's spent grains. *Antioxidants* **2019**, *8*, 380. [CrossRef] [PubMed]

7. Matos, M.; Romero-Díez, R.; Álvarez, A.; Bronze, M.; Rodríguez-Rojo, S.; Mato, R.; Cocero, M.; Matias, A. Polyphenol-rich extracts obtained from winemaking waste streams as natural ingredients with cosmeceutical potential. *Antioxidants* **2019**, *8*, 355. [CrossRef] [PubMed]

8. Martín-García, B.; Pasini, F.; Verardo, V.; Díaz-de-Cerio, E.; Tylewicz, U.; Gómez-Caravaca, A.; Caboni, M. Optimization of sonotrode ultrasonic-assisted extraction of proanthocyanidins from brewers' spent grains. *Antioxidants* **2019**, *8*, 282. [CrossRef] [PubMed]

9. Abi-Khattar, A.; Rajha, H.; Abdel-Massih, R.; Maroun, R.; Louka, N.; Debs, E. Intensification of polyphenol extraction from olive leaves using Ired-Irrad®, an environmentally-friendly innovative technology. *Antioxidants* **2019**, *8*, 227. [CrossRef] [PubMed]

10. Fernandes, P.; Ferreira, S.; Bastos, R.; Ferreira, I.; Cruz, M.; Pinto, A.; Coelho, E.; Passos, C.; Coimbra, M.; Cardoso, S.; et al. Apple pomace extract as a sustainable food ingredient. *Antioxidants* **2019**, *8*, 189. [CrossRef] [PubMed]

11. Li, Y.; Bundeesomchok, K.; Rakotomanomana, N.; Fabiano-Tixier, A.; Bott, R.; Wang, Y.; Chemat, F. Towards a zero-waste biorefinery using edible oils as solvents for the green extraction of volatile and non-volatile bioactive compounds from rosemary. *Antioxidants* **2019**, *8*, 140. [CrossRef] [PubMed]

12. Lakka, A.; Grigorakis, S.; Karageorgou, I.; Batra, G.; Kaltsa, O.; Bozinou, E.; Lalas, S.; Makris, D. Saffron processing wastes as a bioresource of high-value added compounds: Development of a green extraction process for polyphenol recovery using a natural deep eutectic solvent. *Antioxidants* **2019**, *8*, 586. [CrossRef] [PubMed]

 *antioxidants*

Article

Saffron Processing Wastes as a Bioresource of High-Value Added Compounds: Development of a Green Extraction Process for Polyphenol Recovery Using a Natural Deep Eutectic Solvent

Achillia Lakka [1], Spyros Grigorakis [2], Ioanna Karageorgou [1], Georgia Batra [1], Olga Kaltsa [1], Eleni Bozinou [1], Stavros Lalas [1] and Dimitris P. Makris [1,*]

[1] School of Agricultural Sciences, University of Thessaly, N. Temponera Street, 43100 Karditsa, Greece; ACHLAKKA@uth.gr (A.L.); ikarageorgou@uth.gr (I.K.); gbatra@uth.gr (G.B.); okaltsa@uth.gr (O.K.); empozinou@uth.gr (E.B.); slalas@uth.gr (S.L.)

[2] Food Quality & Chemistry of Natural Products, Mediterranean Agronomic Institute of Chania (M.A.I.Ch.), International Centre for Advanced Mediterranean Agronomic Studies (CIHEAM), P.O. Box 85, 73100 Chania, Greece; grigorakis@maich.gr

* Correspondence: dimitrismakris@uth.gr; Tel.: +30-24410-64792

Received: 31 October 2019; Accepted: 22 November 2019; Published: 25 November 2019

Abstract: The current investigation was undertaken to examine saffron processing waste (SPW) as a bioresource, which could be valorized to produce extracts rich in antioxidant polyphenols, using a green, natural deep eutectic solvent (DES). Initially, there was an appraisal of the molar ratio of hydrogen bond donor/hydrogen bond acceptor in order to come up with the most efficient DES composed of L-lactic acid/glycine (5:1). The following step was the optimization of the extraction process using response surface methodology. The optimal conditions thus determined were a DES concentration of 55% (*w/v*), a liquid-to-solid ratio of 60 mL g^{-1}, and a stirring speed of 800 rounds per minute. Under these conditions, the extraction yield in total polyphenols achieved was 132.43 ± 10.63 mg gallic acid equivalents per g of dry mass. The temperature assay performed within a range of 23 to 80 °C, suggested that extracts displayed maximum yield and antioxidant activity at 50–60 °C. Liquid chromatography-mass spectrometry analysis of the SPW extract obtained under optimal conditions showed that the predominant flavonol was kaempferol 3-*O*-sophoroside and the major anthocyanin delphinidin 3,5-di-*O*-glucoside. The results indicated that SPW extraction with the DES used is a green and efficient methodology and may afford extracts rich flavonols and anthocyanins, which are considered to be powerful antioxidants.

Keywords: anthocyanins; antioxidants; deep eutectic solvents; extraction; polyphenols; saffron

---

1. Introduction

In recent years, the agri-food sector has been acknowledged as a major contributor to the global environmental burden. Processing of plants (fruit, vegetables, tubers etc.) for the production of plant food commodities is considered to be a major concern, since a vast amount of waste material may be generated [1]. Plant processing waste is residual biomass rich in moisture and microbial loads and can be a direct risk associated with environmental pollution. On the other hand, an ever-increasing number of current studies on plant food processing residues suggests the presence of a wide range of bioactive compounds in different waste fractions. These bioactive substances are primarily secondary plant metabolites, belonging to polyphenols, carotenoids, essential oils, resins, etc. Therefore, plant food processing waste and residues are highly regarded as very promising sources of bioactive compounds, with applications in food technology, pharmaceuticals, and cosmetics [2,3].

To date, the development of methodologies for high-performance and time-effective extraction of polyphenols from plant matrices is a challenge, because of the inherent limitations of conventional extraction methods. The valorization of polyphenols as bioactive ingredients at various commercial levels has shifted research to low-cost, eco-friendly, and efficient extraction techniques, based on a green philosophy [4]. A basic concept of such an approach would be the use of novel, green solvents, which would be devoid of the disadvantages that characterize the conventional, volatile, petroleum-based solvents. In this view, the emerging liquids known as deep eutectic solvents (DES) would appear to be solid ground for the implementation of green processes for the production of polyphenol-enriched extracts.

DES are novel materials, which can be synthesized using natural substances, such as sugars, polyols, organic acids and their salts, amino acids, etc. [5]. They are usually composed of a hydrogen bond donor (HBD) and a hydrogen bond acceptor (HBA). The ongoing research on these solvents has provided substantial evidence that they can be highly effective in polyphenol extraction, surpassing the potency of conventional solvents, such as methanol. On the other hand, DES are not volatile, their production does not depend on fossil sources, and they have very attractive characteristics, including tunability of composition (and thus regulation of their properties), lack of toxicity, recyclability, and low cost. It is not surprising, therefore, that over the past five years, numerous DES have been synthesized and tested for their potency to extract polyphenolic compounds [6,7].

The plant *Crocus sativus* (Iridaceae), known widely as saffron, is a perennial herb that has been acknowledged since antiquity for its culinary uses and medicinal properties [8]. The most precious part of the plant is the stigmas, which are collected and dried to produce the world's most expensive spice. Following screening and separation, the rest of the flower, composed essentially of the tepals (undifferentiated petals and sepals), is rejected as a residual material. However, emerging evidence has showed that saffron petals contain an array of bioactive polyphenols, including a series of flavonol glycosides and anthocyanin pigments. Several of these constituents were reported to possess multiple beneficial bioactivities [9], and on this evidence, a few extraction methodologies were developed, with the aim of producing polyphenol-containing extracts from saffron processing waste (SPW) [10–13].

However, to the best of the authors' knowledge, the use of DES has never been reported for SPW extraction. The present investigation describes the development of a green extraction methodology for the effective recovery of SPW polyphenols, using a DES composed of L-lactic acid (HBD) and glycine (HBA). The study included the synthesis of the most efficient system by screening a range of HBD:HBA molar ratios and then the optimization by deploying response surface methodology and a temperature assay. The polyphenolic composition of the optimally obtained extract was assessed by performing liquid chromatography, mass spectrometry analyses.

2. Materials and Methods

2.1. Chemicals

Glycine (99.5%) was from Applichem (Darmstadt, Germany). Iron chloride hexahydrate was from Merck (Darmstadt, Germany). Rutin (quercetin 3-*O*-rutinoside) hydrate, kaempferol 3-*O*-glucoside, 2,4,6-tris(2-pyridyl)-*s*-triazine (TPTZ), 2,2-diphenylpicrylhydrazyl (DPPH) and Folin-Ciocalteu reagent were from Sigma-Aldrich (St. Louis, MO, USA). L-Lactic acid, sodium carbonate anhydrous (99%), ascorbic acid (99.5%), and sodium acetate trihydrate and aluminium chloride anhydrous (98%) were from Penta (Praha, Czechia). Gallic acid hydrate was from Panreac (Barcelona, Spain). Pelargonin (pelargonidin 3,5-di-*O*-glucoside) chloride was from Extrasynthese (Genay, France).

2.2. Plant Material and Handling

Saffron (*Crocus sativus* L.) processing waste (SPW), composed essentially of saffron tepals, was collected immediately after manual processing of saffron flowers from a processing plant located in Kozani (West Macedonia, Greece). The plant material was transferred to the laboratory within 24 h and dried at 55 °C for 48 h in a laboratory oven (Binder BD56, Bohemia, NY, USA). Dried SPW was pulverized in a ball-mill to give powders with approximate average particle diameter of 0.317 mm, and stored in air-tight vessels at −18 °C until used.

2.3. DES Synthesis

Synthesis of the DES used in this study was based on a previous protocol [14]. Exact weights of L-lactic acid (HBD) and glycine (HBA) were transferred into a round-bottom glass flask and heated moderately (75–80 °C) for approximately 120 min until the formation of a perfectly transparent liquid. Heating was provided by an oil bath placed on a thermostat-equipped hotplate (Witeg, Wertheim, Germany). The liquid was allowed to acquire room temperature and stored in a sealed vial, in the dark. Inspection for appearance of crystals that would indicate instability was performed at regular intervals over six weeks.

2.4. Batch Stirred-Tank Extraction

Exact mass of 0.570 g of dried plant material was introduced into a 50-mL round-bottom flask with 20 mL of solvent to give a liquid-to-solid ratio ($R_{L/S}$) of 35 mL g^{-1}. The flask was immersed into oil bath and heated by means of a thermostat-equipped hotplate. Extractions were carried out for 150 min, at 50 °C, under magnetic stirring set at 500 rpm. All DES were tested as 70% (*w/v*) aqueous mixtures. Extractions with deionised water, 60% (*v/v*) aqueous ethanol and 60% (*v/v*) aqueous methanol were used as control. After the extraction, samples were centrifuged at 10,000× *g* for 10 min and the supernatant was used for all analyses.

2.5. Extraction Optimization with Response Surface Methodology (RSM)

The scope of RSM was the implementation of a mathematical model to predict polyphenol extraction performance from SPW using the most efficient DES synthesized. The mode chosen was a Box-Behnken experimental design with three central points. Key extraction variables including the concentration of DES in aqueous mixtures (C_{DES}), the liquid-to-solid ratio ($R_{L/S}$) and the stirring speed (S_S) [15] were taken into account and termed X_1, X_2, and X_3, respectively (Table 1). Yield in total polyphenols (Y_{TP}) was the screening response and the three independent variables were coded between −1 (lower limit) and 1 (upper limit). Codification was performed with the following equation [16]:

$$X_i = \left(\frac{z_i - z_1^0}{\Delta z_i}\right) \times \beta_d.$$

(1)

Table 1. Codified and actual values of the independent variables considered for the experimental design.

Independent Variables	Code Units	Coded Variable Level		
		−1	−1	−1
C_{DES} (%, *w/v*)	X_1	55	70	85
$R_{L/S}$ (mL g^{-1})	X_2	20	40	60
S_S (rpm)	X_3	200	500	800

Δz_i is the distance between the real value at the central design point and the real value in the upper or lower limit of a variable; β_d is the major coded limit value in the matrix for each variable, and z^0 is the real value at the central point. The equation (mathematical model) obtained by fitting the function to the experimental data was evaluated by ANOVA. Visual model representation was done by 3D surface response plots.

2.6. Total Polyphenol Determination

An established methodology was used [17]. Samples were diluted 1:50 with 0.5% aqueous formic acid prior to determinations. A volume of 0.1 mL of diluted sample was transferred into a 1.5-mL Eppendorf tube and mixed with 0.1 mL Folin–Ciocalteu reagent. The mixture was allowed to react for 2 min and then 0.8 mL of sodium carbonate (5% *w/v*) was added, followed by 20-min incubation at 40 °C, in a water bath. After incubation, the absorbance at 740 nm was read and total polyphenol concentration (C_{TP}) was determined from a calibration curve constructed with gallic acid (10–80 mg L^{-1}). Extraction yield in total polyphenols was expressed as mg gallic acid equivalents (GAE) per g dry mass (dm).

2.7. Total Flavonoid Determination

For total flavonoids, a previously published protocol was employed [18]. Volume of 0.1 mL of appropriately diluted sample was combined with 0.86 mL 35% (*v/v*) aqueous ethanol and 0.04 mL of reagent consisted of 5% (*w/v*) $AlCl_3$ and 0.5 M CH_3COONa. After 30 min at room temperature the absorbance was obtained at 415 nm. Total flavonoid concentration (C_{TFn}) was calculated from a calibration curve using rutin as standard (15–300 mg L^{-1}). Yield in total flavonoids (Y_{TFn}) was estimated as mg rutin equivalents (RtE) per g dm.

2.8. Determination of the Antiradical Activity (A_{AR})

The determination was based on the stable radical probe DPPH using a stoichiometric assay [19]. All samples were diluted 1:50 with methanol just before the analysis, and 0.025 mL of sample was mixed with 0.975 mL DPPH (100 µM in methanol) at room temperature. Absorbance readings at 515 nm were performed at $t = 0$ min (immediately after mixing) and at $t = 30$ min. The A_{AR} of the extract was then computed as follows:

$$A_{AR} = \frac{C_{DPPH}}{C_{TP}} \times (1 - \frac{A_{515(f)}}{A_{515(i)}}) \times Y_{TP}. \tag{2}$$

C_{DPPH} and C_{TP} are the DPPH concentration (µM) and total polyphenol concentration (mg L^{-1}) in the reaction mixture, respectively. $A_{515(f)}$ corresponds to A_{515} at $t = 30$ min and $A_{515(i)}$ to A_{515} at $t = 0$. Y_{TP} is the extraction yield (mg g^{-1}) in TP of each of the extracts tested. A_{AR} was calculated as µmol DPPH g^{-1} dm.

2.9. Determination of the Reducing Power (P_R)

The ferric-reducing power assay was performed as previously described [19]. Before the analysis, samples were diluted 1:50. Then, 0.05 mL of the sample was incubated with 0.05 mL $FeCl_3$ (4 mM in 0.05 M HCl) at 37 °C in a water bath for 30 min. Following incubation, 0.9 mL of TPTZ solution (1 mM in 0.05 M HCl) was added, and the mixture was allowed to stand at room temperature for further 5 min. The absorbance was obtained at 620 nm and P_R was reported as µmol ascorbic acid equivalents (AAE) g^{-1} dm using an ascorbic acid calibration curve (50–300 µM).

2.10. Liquid Chromatography Diode Array Mass Spectrometry (LC-DAD-MS)

A modification of a method reported elsewhere was used [20]. The apparatus was a Finnigan (San Jose, CA, USA) MAT Spectra System P4000 pump, a UV6000LP diode array detector, and a Finnigan AQA mass spectrometer. Analyses were performed with a Fortis RP-18 column, 150 mm × 2.1 mm, 3 µm, at 40 °C, with a 10-µL injection loop. Acquisition of mass spectra at 20 and 70 eV was performed with electrospray ionization (ESI) in positive ion mode, using the following settings: probe temperature was 250 °C, source voltage at 25 V, detector voltage at 450 V, and capillary voltage at 4 kV. The eluents were (A) 2% acetic acid and (B) methanol and the flow rate was 0.3 mL min^{-1}. Elution was carried out as follows: 0–30 min, 0–100% methanol; 30–40 min, 100% methanol.

2.11. High-Performance Liquid Chromatography Diode Array (HPLC-DAD)

The analysis was carried out on a Shimadzu CBM-20A liquid chromatograph (Shimadzu Europa GmbH, Duisburg, Germany) equipped with an SIL-20AC auto sampler and a CTO-20AC column oven. Detection was carried out using a Shimadzu SPD-M20A detector. The system was interfaced by Shimadzu LC solution software. Chromatography was carried out on a Phenomenex Luna C18(2) column (100 Å, 5 μm, 4.6 × 250 mm) (Phenomenex, Inc., Torrance, CA, USA). Columns were maintained at a temperature of 40 °C. Eluents were (A) 0.5% aqueous formic acid and (B) 0.5% formic acid in MeCN/water (6:4), and the flow rate was 1 mL min^{-1}. A 20 μL sample was injected into the high-performance liquid chromatography (HPLC). Following is the elution program used: 100% A to 60% A in 40 min, 60% A to 50% A in 10 min, 50% A to 30% A in 10 min, and then isocratic elution for another 10 min. The column was washed with 100% MeCN and re-equilibrated with 100% eluent A before the next injection. Quantification was performed with calibration curves (0–50 μg mL^{-1}) constructed with kaempferol 3-*O*-glucoside ($R^2 = 0.9999$), rutin ($R^2 = 0.9990$), and pelargonin ($R^2 = 0.9999$).

2.12. Statistical Analysis

Extractions were repeated at least twice, and all determinations were carried out in triplicate. Values presented are means ± standard deviation (sd). Linear regression analysis was used to establish linear correlations, at least at a 95% significance level ($p < 0.05$), using SigmaPlot™ 12.5 (Systat Software Inc., San Jose, CA, USA). The design of experiment, response surface methodology and all associated statistics were performed with JMP™ Pro 13 (SAS, Cary, NC, USA).

3. Results and Discussion

3.1. DES Synthesis and the Effect of HBD:HBA Molar Ratio ($R_{mol}^{D/A}$)

The first report on L-lactic acid (LA) and glycine (Gly) combination pointed out that stable DES may be formed at $R_{mol}^{D/A} > 3$ [21]. In latter studies, DES composed of LA and Gly exhibited stability at $R_{mol}^{D/A} \geq 5$, and it was also demonstrated that $R_{mol}^{D/A}$ may significantly affect DES efficiency in extracting phenolics [22]. On such a basis, synthesis and screening of a series of LA-Gly DES with $R_{mol}^{D/A}$ ranging from five to 13, was the first stage in the development of an efficient solvent. All DES synthesized were tested for polyphenol recovery as 70% (*w/v*) aqueous mixtures. Screening results are depicted in Figure 1. The DES LA-Gly (5:1) was proven to be the highest-performing system, providing significantly increased Y_{TP} ($p < 0.05$). This finding evidenced the potency of LA-Gly (5:1) for polyphenol recovery, and on this ground, this DES was chosen for all further processes.

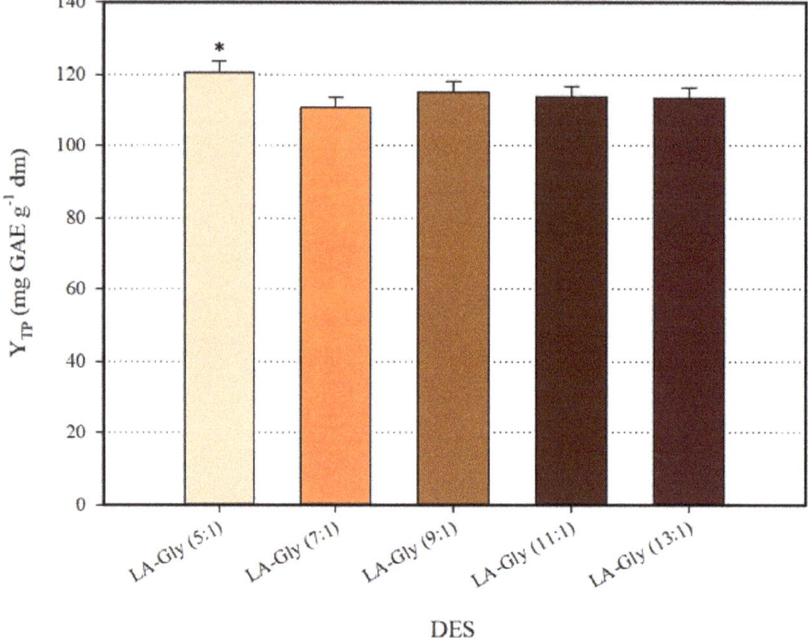

Figure 1. Screening of HBD:HBA ratio for the selection of the most efficient deep eutectic solvent (DES). Asterisk (*) denotes statistically different value ($p < 0.05$).

3.2. Assessment of the DES Extraction Efficiency

To better illustrate the efficiency of LA-Gly (5:1), an appraisal was carried out by comparing the DES performance with that of two other green solvents, namely 60% (v/v) aqueous ethanol and water. Extractions with a commonly used solvent, 60% (v/v) aqueous methanol, were also performed. For the appraisal, in addition to Y_{TP}, the Y_{TFn}, A_{AR} and P_R were also considered, and the results are analytically displayed in Table 2. LA-Gly (5:1) gave higher Y_{TP}, which was statistically significant ($p < 0.05$). Regarding Y_{TFn} and P_R, extraction of SPW with LA-Gly (5:1) also afforded higher but statistically non-significant values, whereas A_{AR} of the LA-Gly (5:1) was lower compared to the extracts obtained with the control solvents. Based on these results, it was deemed that LA-Gly (5:1) was indeed the highest-performic system.

Table 2. Extraction yields and antioxidant characteristics of the saffron processing waste (SPW) extracts obtained with DES and the control solvents.

Solvent	Y_{TP} (mg GAE g^{-1} dm)	Y_{TFn} (mg RtE g^{-1} dm)	A_{AR} (μmol DPPH g^{-1} dm)	P_R (μmol AAE g^{-1} dm)
Water	102.91 ± 2.57	49.77 ± 2.99	284.66 ± 5.69	136.14 ± 2.04
60% EtOH	112.15 ± 2.80	53.98 ± 3.24	290.54 ± 5.81	137.18 ± 2.56
60% MeOH	107.13 ± 2.68	54.86 ± 3.29	300.71 ± 6.01	129.05 ± 2.09
DES	120.50 ± 3.01 *	61.27 ± 3.37	213.05 ± 4.26	144.66 ± 3.07

* Asterisk indicates statistically different value ($p < 0.05$).

3.3. Optimisation of Extraction Performance

Response surface methodology was deployed to assess the effect of three basic extraction variables (C_{DES}, $R_{L/S}$, S_S) on the performance of LA-Gly (5:1) to recover polyphenolic antioxidants. The objective was the fit of polynomial equations (models) to the experimental data, in order to describe effectively the behavior of the data set for making statistical previsions. Assessment of the fitted models was based on the ANOVA (Table 3). By neglecting the non-significant terms, the first-degree equation (mathematical model) was

$$Y_{TP} = 115.32 + 3.24X_1 + 8.67X_2 + 5.73X_3 - 5.11X_1X_2 - 4.93X_1X_3. \tag{3}$$

The square correlations coefficient (R^2) and the *p*-value are indicators of the total variability around the mean calculated by the model. Measured and predicted Y_{TP} values for each material extracted and for each design point are analytically presented in Table 4. Because total R^2 of the model was 0.95, and the *p* value (assuming a confidence interval of 95%) was highly significant (0.0080) (*F* value for lack-of-fit = 30.1016), Equation (3) showed excellent fitting to the experimental data.

Table 3. Statistical data related with the model established by implementing response surface methodology.

Term	Standard Error	*t* Ratio	Probability > *t*	Sum of Squares	*F* Ratio
C_{DES}	1.237502	2.62	0.0472 *	83.98080	6.8548
$R_{L/S}$	1.237502	7.01	0.0009 *	601.17781	49.0705
S_S	1.237502	4.63	0.0057 *	262.54861	21.4303
$C_{DES}\,R_{L/S}$	1.750093	−2.92	0.0329 *	104.65290	8.5422
$C_{DES}\,S_S$	1.750093	−2.81	0.0374 *	97.02250	7.9194
$R_{L/S}\,S_S$	1.750093	−1.90	0.1165	44.02322	3.5934
$C_{DES}\,C_{DES}$	1.821554	−1.56	0.1793	29.84188	2.4358
$R_{L/S}\,R_{L/S}$	1.821554	0.64	0.5475	5.09408	0.4158
$S_S\,S_S$	1.821554	0.49	0.6441	2.95488	0.2412

* Asterisk indicates statistically different value ($p < 0.05$).

Table 4. Measured and predicted values of the response for every point of the experimental design implemented.

Design Point	Independent Variables			Response (Y_{TP}, mg GAE g^{-1} dm)	
	X_1 (C_{DES}, % *w/v*)	X_2 ($R_{L/S}$, mL g^{-1})	X_3 (S_S, rpm)	Measured	Predicted
1	−1 (55)	−1 (20)	0 (500)	93.36	96.63
2	−1 (55)	1 (60)	0 (500)	122.14	124.20
3	1 (85)	−1 (20)	0 (500)	115.40	113.34
4	1 (85)	1 (60)	0 (500)	123.72	120.45
5	0 (70)	−1 (20)	−1 (200)	100.43	99.68
6	0 (70)	−1 (20)	1 (800)	118.23	117.77
7	0 (70)	1 (60)	−1 (200)	123.19	123.65
8	0 (70)	1 (60)	1 (800)	127.72	128.47
9	−1 (55)	0 (40)	−1 (200)	102.00	99.48
10	1 (85)	0 (40)	−1 (200)	113.00	115.81
11	−1 (55)	0 (40)	1 (800)	123.60	120.79
12	1 (85)	0 (40)	1 (800)	114.90	117.42
13	0 (70)	0 (40)	0 (500)	116.25	115.32
14	0 (70)	0 (40)	0 (500)	115.00	115.32
15	0 (70)	0 (40)	0 (500)	114.72	115.32

The 3D plots created based on the model are given in Figure 2, to readily portray the effect of the process variables on the response (Y_{TP}). The desirability function enabled the simultaneous optimization of the levels of all three variables in order to attain the best system performance (Figure 3), and the sets of conditions to achieve the highest theoretical yield were estimated to be C_{DES} = 55% (*w/v*), $R_{L/S}$ = 60 mL g^{-1} and S_S = 800 rpm. Under these conditions, the maximum theoretical Y_{TP} was 132.43 ± 10.63 mg GAE g^{-1} dm. To confirm the validity of the model, three extractions of each material were carried out under the optimal conditions. The Y_{TP} determined was 128.00 ± 1.94 mg GAE g^{-1} dm, suggesting that the theoretical optimum settings for C_{DES}, $R_{L/S}$, and S_S may be applied with high reliability.

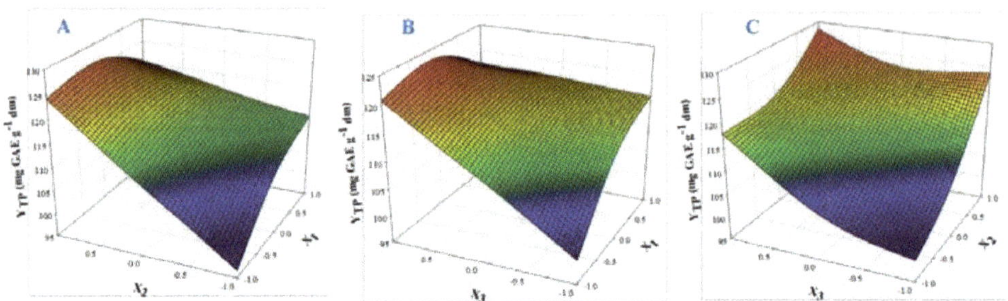

Figure 2. Three-dimensional plots displaying the effect of process (independent) variables on the total polyphenol yield (YTP). For variable assignment, see Table 1. Plots (**A**), (**B**) and (**C**) show covariation of variables X1 and X2, X1 and X3, and X2 and X3, respectively.

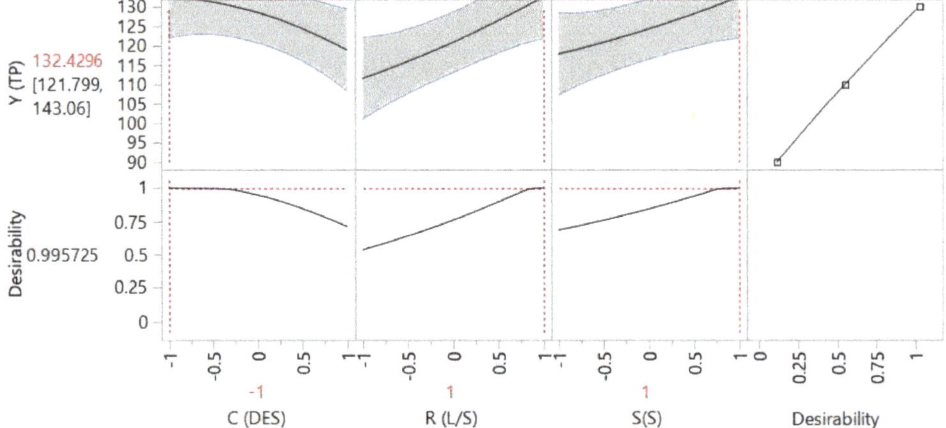

Figure 3. Desirability function showing the maximum predicted response upon setting process (independent) variable values at the predicted optima.

Judging by the model, C_{DES} (X_1) had a direct positive influence on Y_{TP}, and the same was seen for $R_{L/S}$ (X_2). Likewise, S_S (X_3) had a positive impact on SPW polyphenol extraction. On the other hand, cross terms C_{DES} (X_1) and $R_{L/S}$ (X_2), and C_{DES} (X_1) and S_S (X_3) had a negative effect on Y_{TP}. The predicted C_{DES} levels implied the use of a significantly higher water amount compared with previous results from polyphenol extraction with DES, which indicated that 80% (*w/w*) to be the most appropriate C_{DES} for high extraction yield [23–26]. Suitable DES mixing with water is indispensable for regulation of properties crucial to solid-liquid extraction, such as viscosity and polarity [27]. However, C_{DES} cannot be below a certain level because excessive water amount would cause DES decomposition and therefore the intrinsic DES properties would be abolished [28].

Variable $R_{L/S}$ is a strongly influential factor regarding solid–liquid extraction, as it affects concentration gradient between the solid particles and the liquid phase, which is the driving force for the manifestation of diffusion phenomena. Recently, it was demonstrated that raising $R_{L/S}$ from 10 to 50 mL g^{-1} may significantly increase diffusivity [29]. Conventional solvent extraction may require $R_{L/S}$ as high as 120 mL g^{-1} [30,31], but for polyphenol extractions with DES, lower $R_{L/S}$ levels ranging from 29–50 mL g^{-1} are usually effective [32–34]. The optimal $R_{L/S}$ estimated was 60 mL g^{-1}, indicating that higher concentration gradients may be necessary for effective polyphenol leaching into the liquid phase.

In a similar manner, S_S was shown to play an important role in solid–liquid extraction, and appropriate S_S regulation may give significantly higher extraction yields [29,31]. Sufficiently high S_S causes turbulence in the extraction tank, and this in turn may increase mass transfer rate. In this line, S_S has been demonstrated to provide increased polyphenol diffusivity [29]. On the other hand, high S_S may lead to incomplete diffusivity because higher turbulence could shift the equilibrium toward polyphenol adsorption rather than diffusion. At this point, characteristics such as the viscosity of the liquid phase (solvent), which is tightly associated with $R_{L/S}$, should also be considered. Such a hypothesis might explain the combined effect observed between S_S and C_{DES} (cross term X_1X_3) for SPW polyphenol extraction.

3.4. Temperature Effects

Polyphenols are thermosensitive molecules and in several cases temperature increase does not generate a monotonous effect on the extraction yield and antioxidant activity. Such a behavior was demonstrated for the extraction of onion solid waste [35,36], red grape pomace [37], and *Moringa oleifera* leaves [25]. Therefore, the impact of temperature on the production of polyphenol-enriched extracts with improved antioxidant characteristics merits thorough investigation. For this reason, extractions under optimal conditions were carried out at temperatures varying from 23 (ambient temperature) to 80 °C, and the extracts produced were assessed by determining Y_{TP}, Y_{TFn}, A_{AR}, and P_R. The outcome of this assay is analytically presented in Figure 4.

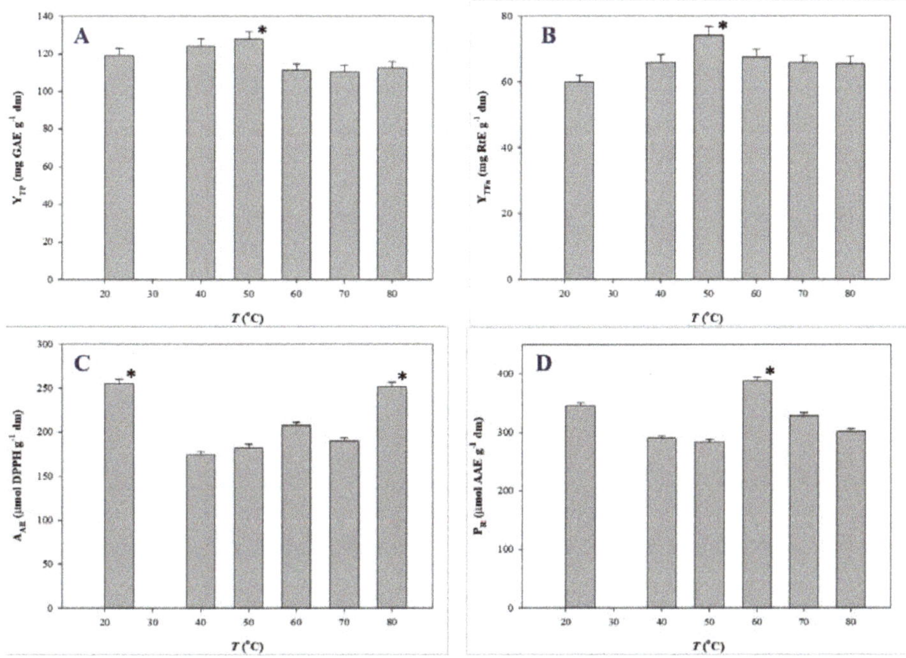

Figure 4. Variation of Y_{TP} (**A**), Y_{TFn} (**B**), A_{AR} (**C**), and P_R (**D**) as a function of T. Bars indicate standard deviation. Asterisk (*) denotes statistically different value ($p < 0.05$).

Y_{TP} for SPW extraction peaked at 50 °C (128.03 mg GAE g^{-1} dm), while a decline was recorded thereafter (Figure 4A). Likewise, maximum Y_{TFn} was found at 50 °C (Figure 4B). A_{AR} exhibited fluctuations within a narrow range, the highest A_{AR} being at 23 °C (255.15 µmol DPPH g^{-1} dm) (Figure 4C). Significant differentiation was seen for the evolution of P_R, which gave maximum levels at

60 °C (388.15 µmol AAE g^{-1} dm) (Figure 4D) Considering all the above parameters, the data obtained suggested that SPW extraction provided polyphenol-enriched extracts with enhanced antioxidant activity at around 50–60 °C. This finding may be evidence of the thermal stability of SPW constituents.

3.5. Polyphenolic Composition

The extract obtained under optimal conditions (C_{DES} = 55% (*w/v*), $R_{L/S}$ = 60 mL g^{-1}, S_S = 800 rpm, *T* = 50 °C) was analyzed by liquid chromatography-diode array-mass spectrometry, to detect and tentatively identify the major polyphenolic constituents. The compound with retention time (Rt) 15.62 min (peak #5) in the chromatogram monitored at 520 nm (Figure 5), gave a molecular ion at *m/z* = 627 and a diagnostic fragment at *m/z* = 465. This peak was identified as delphinidin 3,5-di-*O*-glucoside (Table 5). Likewise, the peak at 18.72 min (peak #6) yielded a molecular ion at *m/z* = 641 and a characteristic fragment at *m/z* = 465, and it was assigned to petunidin 3,5-di-*O*-glucoside. Finally, the peak with Rt 20.08 min (peak #7) was identified as delphinidin 3-*O*-glucoside, based on its major peak (*m/z* = 465) and its fragment at *m/z* = 303 [38].

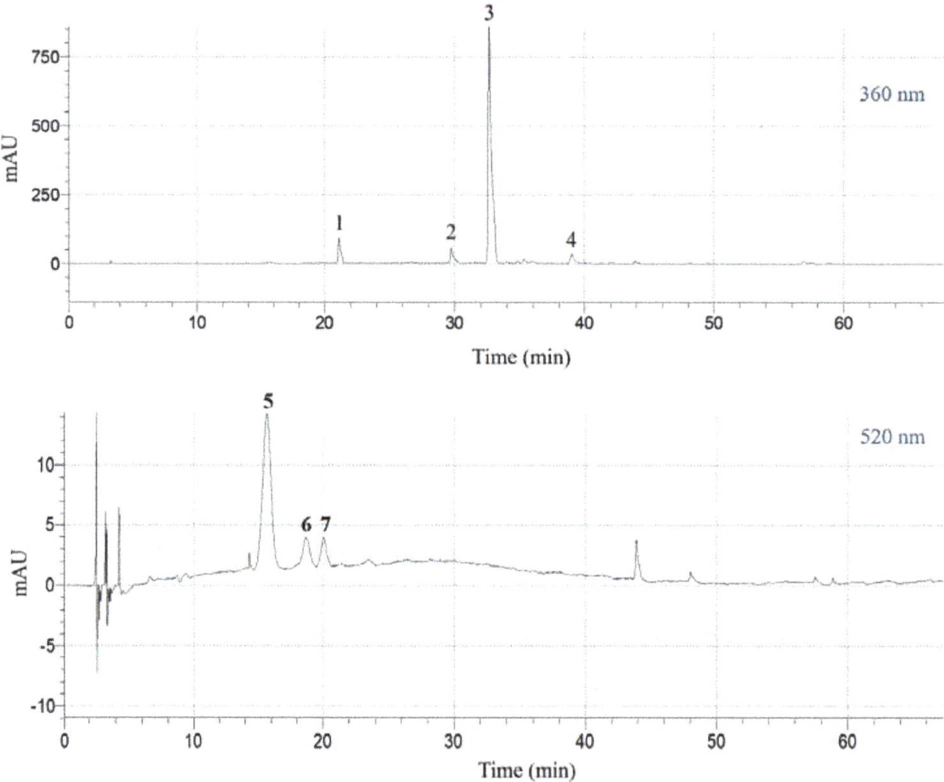

Figure 5. High-performance liquid chromatography (HPLC) traces of the SPW extract obtained with the DES, under optimal conditions (C_{DES} = 55% (*w/v*), $R_{L/S}$ = 60 mL g^{-1}, S_S = 800 rpm, *T* = 50 °C). The upper and lower traces were monitored at 360 and 520 nm, respectively. Peak assignment: 1, kaempferol 3-*O*-sophoroside 7-*O*-glucoside; 2, quercetin 3-*O*-sophoroside; 3, kaempferol 3-*O*-sophoroside; 4, kaempferol 3-*O*-glucoside; 5, delphinidin 3,5-di-*O*-glucoside; 6, petunidin 3,5-di-*O*-glucoside; 7, delphinidin 3-*O*-glucoside.

The chromatogram at 360 nm revealed the existence of four principal constituents (Figure 5). Peak #1 yielded a pseudo-molecular ion at *m/z* = 773 and three diagnostic fragments at *m/z* = 611 (loss of glucose) at $m/z = 449$ (loss of sophorose) and at $m/z = 287$ (aglycone). This compound was tentatively identified as kaempferol 3-*O*-sophoroside 7-*O*-glucoside (Table 5). Similarly, peak #2 gave a pseudo-molecular ion at $m/z = 627$ and the aglycone ion at $m/z = 303$ and it was assigned to quercetin 3-*O*-sophoroside. Peak #3 displayed a pseudo-molecular ion at $m/z = 611$ and fragments at $m/z = 449$ (loss of glucose) and $m/z = 287$ (aglycone), and its structure was assigned to kaempferol 3,7-di-*O*-glucoside. Peak #4 showed pseudo-molecular and fragment ions at $m/z = 449$ and 287, respectively, and it was identified as kaempferol 3-*O*-glucoside [39].

On the ground of the quantitative data presented in Table 6, the flavonol composition of the extract was characterized by relatively high amounts of kaempferol 3-*O*-sophoroside (36.43 ± 2.55 mg g^{-1} dm), accompanied by much lower proportions of kaempferol 3-*O*-sophoroside 7-*O*-glucoside and quercetin 3-*O*-sophoroside. Regarding anthocyanins, the profile was dominated by delphinidin 3,5-di-*O*-glucoside (6.28 ± 0.44 mg g^{-1} dm). This outcome is in accordance with previous studies [40,41], which showed that the predominant flavonol found in aqueous SPW extract was kaempferol 3-*O*-sophoroside (30.34 mg g^{-1} dm), followed by kaempferol 3-*O*-sophoroside 7-*O*-glucoside (5.6 mg g^{-1} dm) and quercetin 3-*O*-sophoroside (4.01 mg g^{-1} dm). However, important amounts of delphinidin 3,5-di-*O*-glucoside (23.19 mg g^{-1} dm) were determined in ethanolic extract, whereas the aqueous extract was relatively rich in petunidin 3,5-diglucoside (3.97 mg g^{-1} dm). Data from another investigation were in line, giving values of 12.60 and 3.94 mg g^{-1} dm, for kaempferol 3-*O*-sophoroside and delphinidin 3,5-di-*O*-glucoside, respectively, in water/methanol extracts [38]. Although the polyphenolic composition of SPW could be influenced by the genetic background (variety), the area of origin and sample processing conditions, comparison of the total polyphenol content determined in this study (Table 6) with the values reported in the literature would indicate that SPW extraction with the DES used, under the optimal conditions estimated, an efficient process to produce polyphenol-enriched extracts with important antioxidant activity.

Table 5. Ultraviolet-visual and mass spectrometric data of the major polyphenols detected in the DES extracts of SPW, obtained under optimal conditions.

No	Rt (min)	UV-vis	[M + H]$^+$ (m/z)	Fragment Ions (m/z)	Tentative Identity
Flavonols					
1	21.08	265, 346	773	611, 449, 287	Kaempferol 3-*O*-sophoroside 7-*O*-glucoside
2	29.81	254, 351	627	303	Quercetin 3-*O*-sophoroside
3	32.63	265, 346	611	449, 287	Kaempferol 3-*O*-sophoroside
4	38.87	265, 352	449	287	Kaempferol 3-*O*-glucoside
Anthocyanins					
5	15.62	274, 523	627	465	Delphinidin 3,5-di-*O*-glucoside
6	18.72	272, 523	641	465	Petunidin 3,5-di-*O*-glucoside
7	20.08	271, 523	465	303	Delphinidin 3-*O*-glucoside

Table 6. Quantitative values of the major polyphenols detected in the SPW, obtained with the DES under optimal conditions.

Polyphenol	Content (mg g^{-1} dm) $\pm$ sd
Flavonols	
Kaempferol 3-*O*-sophoroside 7-*O*-glucoside	3.92 $\pm$ 0.27
Quercetin 3-*O*-sophoroside	3.55 $\pm$ 0.25
Kaempferol 3-*O*-sophoroside	36.43 $\pm$ 2.55
Kaempferol 3-*O*-glucoside	1.82 $\pm$ 0.13
Total	*45.72*
Anthocyanins	
Delphinidin 3,5-di-*O*-glucoside	6.28 $\pm$ 0.44
Petunidin 3,5-di-*O*-glucoside	1.08 $\pm$ 0.08
Delphinidin 3-*O*-glucoside	0.70 $\pm$ 0.05
Total	*8.06*
Sum	*53.79*

4. Conclusions

Saffron processing waste was used as raw material for the recovery of antioxidant polyphenols using a natural deep eutectic solvent composed of L-lactic acid and glycine. It was demonstrated that the HBD:HBA molar ratio can significantly affect extraction yield, hence initial screening of the most appropriate HBD:HBA molar ratio should be a key step in the development of similar processes. Furthermore, the response surface optimization of polyphenol extraction from SPW clearly showed that the proportion of solvent/water, as well as the liquid-to-solid ratio and the stirring speed may crucially affect the extraction performance. Therefore, these variables are to be suitably adjusted in order to maximize extraction yield. Likewise, temperatures higher than 50–60 °C were shown to have a negative impact on the extraction yield, and this is another salient process parameter that should be taken into consideration. SPW extraction under the optimally defined conditions gave extracts rich in polyphenols, the predominant flavonol and anthocyanin being kaempferol 3-*O*-sophoroside and delphinidin 3,5-di-*O*-glucoside, respectively. Future work should focus on the stability of SPW extracts in DES, as well as on the bioactivity of the extracts, to fully evaluate their potency as food and cosmetic ingredients and pharmaceutical formulations.

Author Contributions: Conceptualization, S.L. and D.P.M.; formal analysis, A.L., S.G., I.K., O.K., G.B., and E.B.; investigation, A.L., S.G., I.K., O.K., G.B., and E.B.; writing—original draft preparation, S.L. and D.P.M.; writing—review and editing, S.L. and D.P.M.; supervision, S.L. and D.P.M.

Funding: This research was co-financed by the European Union and the Hellenic Ministry of Economy and Development through the Operational Programme Competitiveness, Entrepreneurship and Innovation, under the call RESEARCH—CREATE—INNOVATE (project code: T1EDK-05677).

Acknowledgments: Nektarios Vlahos (Petrana, Kozani) is thanked for providing the saffron processing waste.

Conflicts of Interest: The authors declare no conflict of interest.

Nomenclature

A_{AR}	antiradical activity (μmol DPPH g^{-1})
P_R	reducing power (μmol AAE g^{-1})
$R_{L/S}$	liquid-to-solid ratio (mL g^{-1})
t	time (min)
T	temperature (°C)
Y_{TFn}	yield in total flavonoids (mg RtE g^{-1})
Y_{TP}	yield in total polyphenols (mg GAE g^{-1})

Abbreviations

AAE	ascorbic acid equivalents
DES	deep eutectic solvents
DPPH	2,2-diphenyl-1-picrylhydrazyl radical
GAE	gallic acid equivalents
HBA	hydrogen bond acceptor
HBD	hydrogen bond donor
TPTZ	2,4,6-tripyridyl-*s*-triazine

References

1. Banerjee, J.; Singh, R.; Vijayaraghavan, R.; MacFarlane, D.; Patti, A.F.; Arora, A. Bioactives from fruit processing wastes: Green approaches to valuable chemicals. *Food Chem.* **2017**, *225*, 10–22. [CrossRef] [PubMed]
2. Makris, D.P.; Boskou, D. Plant-Derived Antioxidants as Food Additives. In *Plants as a Source of Natural Antioxidants*; CABI Publ.: Oxfordshire, UK, 2014; pp. 169–190.
3. Shahidi, F.; Varatharajan, V.; Oh, W.Y.; Peng, H. Phenolic compounds in agri-food by-products, their bioavailability and health effects. *J. Food Bioact* **2019**, *5*, 57–119. [CrossRef]
4. Chemat, F.; Abert Vian, M.; Ravi, H.K.; Khadhraoui, B.; Hilali, S.; Perino, S.; Tixier, A.S.F. Review of alternative solvents for green extraction of food and natural products: Panorama, principles, applications and prospects. *Molecules* **2019**, *24*, 3007. [CrossRef] [PubMed]
5. Florindo, C.; Lima, F.; Ribeiro, B.D.; Marrucho, I.M. Deep eutectic solvents: Overcoming XXI century challenges. *Cur. Opin. Green Sustain. Chem.* **2018**. [CrossRef]
6. Bubalo, M.C.; Vidović, S.; Redovniković, I.R.; Jokić, S. New perspective in extraction of plant biologically active compounds by green solvents. *Food Bioprod. Process.* **2018**, *109*, 52–73. [CrossRef]
7. Li, Y.; Kunz, W.; Chemat, F. From Petroleum to Bio-Based Solvents: From Academia to Industry. In *Plant Based "Green Chemistry 2.0"*; Springer: Berlin/Heidelberg, Germany, 2019; pp. 51–87.
8. Ghaffari, S.; Roshanravan, N. Saffron; An updated review on biological properties with special focus on cardiovascular effects. *Biomed. Pharmacother.* **2019**, *109*, 21–27. [CrossRef]
9. Moratalla-López, N.; Bagur, M.J.; Lorenzo, C.; Salinas, M.; Alonso, G.L. Bioactivity and bioavailability of the major metabolites of *Crocus sativus* L. Flower. *Molecules* **2019**, *24*, 2827. [CrossRef]
10. Ahmadian-Kouchaksaraie, Z.; Niazmand, R.; Najafi, M.N. Optimization of the subcritical water extraction of phenolic antioxidants from *Crocus sativus* petals of saffron industry residues: Box-Behnken design and principal component analysis. *Innov. Food Sci. Emerg. Technol.* **2016**, *36*, 234–244. [CrossRef]
11. Da Porto, C.; Natolino, A. Extraction kinetic modelling of total polyphenols and total anthocyanins from saffron floral bio-residues: Comparison of extraction methods. *Food Chem.* **2018**, *258*, 137–143. [CrossRef]
12. Ahmadian-Kouchaksaraie, Z.; Niazmand, R. Supercritical carbon dioxide extraction of antioxidants from *Crocus sativus* petals of saffron industry residues: Optimization using response surface methodology. *J. Supercrit. Fluids* **2017**, *121*, 19–31. [CrossRef]
13. Lotfi, L.; Kalbasi-Ashtari, A.; Hamedi, M.; Ghorbani, F. Effects of enzymatic extraction on anthocyanins yield of saffron tepals (*Crocus sativus*) along with its color properties and structural stability. *J. Food Drug Anal.* **2015**, *23*, 210–218. [CrossRef]
14. Mouratoglou, E.; Malliou, V.; Makris, D.P. Novel glycerol-based natural eutectic mixtures and their efficiency in the ultrasound-assisted extraction of antioxidant polyphenols from agri-food waste biomass. *Waste Biomass Valorization* **2016**, *7*, 1377–1387. [CrossRef]
15. Lakka, A.; Karageorgou, I.; Kaltsa, O.; Batra, G.; Bozinou, E.; Lalas, S.; Makris, D. Polyphenol extraction from *Humulus lupulus* (hop) using a neoteric glycerol/L-alanine deep eutectic solvent: Optimisation, kinetics and the effect of ultrasound-assisted pretreatment. *AgriEngineering* **2019**, *1*, 403–417. [CrossRef]
16. Bezerra, M.A.; Santelli, R.E.; Oliveira, E.P.; Villar, L.S.; Escaleira, L.A. Response surface methodology (RSM) as a tool for optimization in analytical chemistry. *Talanta* **2008**, *76*, 965–977. [CrossRef] [PubMed]
17. Cicco, N.; Lanorte, M.T.; Paraggio, M.; Viggiano, M.; Lattanzio, V. A reproducible, rapid and inexpensive Folin–Ciocalteu micro-method in determining phenolics of plant methanol extracts. *Microchem. J.* **2009**, *91*, 107–110. [CrossRef]

18. Manousaki, A.; Jancheva, M.; Grigorakis, S.; Makris, D. Extraction of antioxidant phenolics from agri-food waste biomass using a newly designed glycerol-based natural low-transition temperature mixture: A comparison with conventional eco-friendly solvents. *Recycling* **2016**, *1*, 194–204. [CrossRef]
19. Paleologou, I.; Vasiliou, A.; Grigorakis, S.; Makris, D.P. Optimisation of a green ultrasound-assisted extraction process for potato peel (Solanum tuberosum) polyphenols using bio-solvents and response surface methodology. *Biomass Convers. Biorefin.* **2016**, *6*, 289–299. [CrossRef]
20. Sadek, E.S.; Makris, D.P.; Kefalas, P. Polyphenolic composition and antioxidant characteristics of kumquat (*Fortunella margarita*) peel fractions. *Plant Foods Hum. Nutr.* **2009**, *64*, 297. [CrossRef]
21. Bakirtzi, C.; Triantafyllidou, K.; Makris, D.P. Novel lactic acid-based natural deep eutectic solvents: Efficiency in the ultrasound-assisted extraction of antioxidant polyphenols from common native Greek medicinal plants. *J. Appl. Res. Med. Aromat. Plants* **2016**, *3*, 120–127. [CrossRef]
22. Kottaras, P.; Koulianos, M.; Makris, D. Low-Transition temperature mixtures (LTTMs) made of bioorganic molecules: Enhanced extraction of antioxidant phenolics from industrial cereal solid wastes. *Recycling* **2017**, *2*, 3. [CrossRef]
23. Cui, Q.; Liu, J.Z.; Wang, L.T.; Kang, Y.F.; Meng, Y.; Jiao, J.; Fu, Y.J. Sustainable deep eutectic solvents preparation and their efficiency in extraction and enrichment of main bioactive flavonoids from sea buckthorn leaves. *J. Clean. Prod.* **2018**, *184*, 826–835. [CrossRef]
24. Huang, Y.; Feng, F.; Jiang, J.; Qiao, Y.; Wu, T.; Voglmeir, J.; Chen, Z.G. Green and efficient extraction of rutin from tartary buckwheat hull by using natural deep eutectic solvents. *Food Chem.* **2017**, *221*, 1400–1405. [CrossRef]
25. Karageorgou, I.; Grigorakis, S.; Lalas, S.; Makris, D.P. Enhanced extraction of antioxidant polyphenols from Moringa oleifera Lam. leaves using a biomolecule-based low-transition temperature mixture. *Eur. Food Res. Technol.* **2017**, *243*, 1839–1848. [CrossRef]
26. Athanasiadis, V.; Grigorakis, S.; Lalas, S.; Makris, D.P. Highly efficient extraction of antioxidant polyphenols from Olea europaea leaves using an eco-friendly glycerol/glycine deep eutectic solvent. *Waste Biomass Valorization* **2018**, *9*, 1985–1992. [CrossRef]
27. Liu, Y.; Friesen, J.B.; McAlpine, J.B.; Lankin, D.C.; Chen, S.N.; Pauli, G.F. Natural deep eutectic solvents: Properties, applications, and perspectives. *J. Nat. Prod.* **2018**, *81*, 679–690. [CrossRef] [PubMed]
28. Dai, Y.; Witkamp, G.J.; Verpoorte, R.; Choi, Y.H. Tailoring properties of natural deep eutectic solvents with water to facilitate their applications. *Food Chem.* **2015**, *187*, 14–19. [CrossRef] [PubMed]
29. Shewale, S.; Rathod, V.K. Extraction of total phenolic content from *Azadirachta indica* or (neem) leaves: Kinetics study. *Prep. Biochem. Biotechnol.* **2018**, *48*, 312–320. [CrossRef]
30. Adjé, F.; Lozano, Y.F.; Lozano, P.; Adima, A.; Chemat, F.; Gaydou, E.M. Optimization of anthocyanin, flavonol and phenolic acid extractions from *Delonix regia* tree flowers using ultrasound-assisted water extraction. *Ind. Crop. Prod.* **2010**, *32*, 439–444. [CrossRef]
31. Vetal, M.D.; Lade, V.G.; Rathod, V.K. Extraction of ursolic acid from Ocimum sanctum leaves: Kinetics and modeling. *Food Bioprod. Process.* **2012**, *90*, 793–798. [CrossRef]
32. Jancheva, M.; Grigorakis, S.; Loupassaki, S.; Makris, D.P. Optimised extraction of antioxidant polyphenols from *Satureja thymbra* using newly designed glycerol-based natural low-transition temperature mixtures (LTTMs). *J. Appl. Res. Med. Aromat. Plants* **2017**, *6*, 31–40. [CrossRef]
33. Dedousi, M.; Mamoudaki, V.; Grigorakis, S.; Makris, D. Ultrasound-assisted extraction of polyphenolic antioxidants from olive (*Olea europaea*) leaves using a novel glycerol/sodium-potassium tartrate low-transition temperature mixture (LTTM). *Environments* **2017**, *4*, 31. [CrossRef]
34. Slim, Z.; Jancheva, M.; Grigorakis, S.; Makris, D.P. Polyphenol extraction from *Origanum dictamnus* using low-transition temperature mixtures composed of glycerol and organic salts: Effect of organic anion carbon chain length. *Chem. Eng. Commun.* **2018**, *205*, 1494–1506. [CrossRef]
35. Khiari, Z.; Makris, D.P.; Kefalas, P. An investigation on the recovery of antioxidant phenolics from onion solid wastes employing water/ethanol-based solvent systems. *Food Bioprocess Technol.* **2009**, *2*, 337. [CrossRef]
36. Katsampa, P.; Valsamedou, E.; Grigorakis, S.; Makris, D.P. A green ultrasound-assisted extraction process for the recovery of antioxidant polyphenols and pigments from onion solid wastes using Box–Behnken experimental design and kinetics. *Ind. Crop. Prod.* **2015**, *77*, 535–543. [CrossRef]

37. Patsea, M.; Stefou, I.; Grigorakis, S.; Makris, D.P. Screening of natural sodium acetate-based low-transition temperature mixtures (LTTMs) for enhanced extraction of antioxidants and pigments from red vinification solid wastes. *Environ. Process.* **2017**, *4*, 123–135. [CrossRef]

38. Goupy, P.; Vian, M.A.; Chemat, F.; Caris-Veyrat, C. Identification and quantification of flavonols, anthocyanins and lutein diesters in tepals of *Crocus sativus* by ultra performance liquid chromatography coupled to diode array and ion trap mass spectrometry detections. *Ind. Crop. Prod.* **2013**, *44*, 496–510. [CrossRef]

39. Tuberoso, C.I.; Rosa, A.; Montoro, P.; Fenu, M.A.; Pizza, C. Antioxidant activity, cytotoxic activity and metabolic profiling of juices obtained from saffron (Crocus sativus L.) floral by-products. *Food Chem.* **2016**, *199*, 18–27. [CrossRef]

40. Cusano, E.; Consonni, R.; Petrakis, E.A.; Astraka, K.; Cagliani, L.R.; Polissiou, M.G. Integrated analytical methodology to investigate bioactive compounds in *Crocus sativus* L. flowers. *Phytochem. Anal.* **2018**, *29*, 476–486. [CrossRef]

41. Serrano-Díaz, J.; Estevan, C.; Sogorb, M.Á.; Carmona, M.; Alonso, G.L.; Vilanova, E. Cytotoxic effect against 3T3 fibroblasts cells of saffron floral bio-residues extracts. *Food Chem.* **2014**, *147*, 55–59. [CrossRef]

Article

Quantitative Analysis of Bioactive Phenanthrenes in *Dioscorea batatas* Decne Peel, a Discarded Biomass from Postharvest Processing

Minyoul Kim [1,2,†], Myeong Ju Gu [1,†], Joon-Goo Lee [3], Jungwook Chin [4], Jong-Sup Bae [5,*] and Dongyup Hahn [1,2,6,*]

1   School of Food Science and Biotechnology, College of Agriculture and Life Sciences, Kyungpook National University, Daegu 41566, Korea; rlaalsduf123@knu.ac.kr (M.K.); rnaudwn@knu.ac.kr (M.J.G.)
2   Department of Integrative Biology, Kyungpook National University, Daegu 41566, Korea
3   Food Standard Division, Ministry of Food and Drug Safety, Cheongju 28159, Korea; capbox@korea.kr
4   New Drug Development Center, Daegu-Gyeongbuk Medical Innovation Foundation, Daegu 41061, Korea; jwchin@dgmif.re.kr
5   College of Pharmacy, CMRI, Research Institute of Pharmaceutical Sciences, BK21 Plus KNU Multi-Omics based Creative Drug Research Team, Kyungpook National University, Daegu 41566, Korea
6   Institute of Agricultural Science and Technology, College of Agriculture and Life Sciences, Kyungpook National University, Daegu 41566, Korea
*   Correspondence: baejs@knu.ac.kr (J.-S.B.); dohahn@knu.ac.kr (D.H.)
†   These authors contributed equally to this work.

Received: 10 October 2019; Accepted: 8 November 2019; Published: 10 November 2019

Abstract: *Dioscorea batatas* Decne (Chinese yam) has been widely cultivated in East Asia for the purposes of food and medicinal uses for centuries. Along with its high nutritional value, the medicinal value of *D. batatas* has been extensively investigated in association with phytochemicals such as allantoin, flavonoids, saponins and phenanthrenes. Phenanthrenes are especially considered the standard marker chemicals of the Chinese yam for their potent bioactivity and availability of analysis with conventional high performance liquid chromatography with ultraviolet detection (HPLC-UV) methods. In order to investigate how much the contents of phenanthrenes are in the actual food products provided for consumers, *D. batatas* tuber was peeled and separated into its peel and flesh as in the conventional processing method. A quantitative analysis using the HPLC-UV method revealed that phenanthrenes are concentrically present in the *D. batatas* peel, while phenanthrenes are present in the flesh under the limit of detection. The difference in the contents of phenanthrenes is estimated to have arisen the considerable difference of antioxidant potential between the peel and the flesh. The results from this study suggest the high value of the discarded biomass of the Chinese yam peel and the necessity for the utilization of the Chinese yam peel.

Keywords: *Dioscorea batatas*; Chinese yam; phenanthrenes; quantitative analysis

---

1. Introduction

Dioscorea batatas Decne (Chinese yam) is a perennial plant widely cultivated across the tropical and subtropical regions of East Asia as a staple or as a traditional medicine [1]. It has high nutritional values because it contains substantial proteins, carbohydrates, vitamins, fats, choline, and indispensable trace elements for the human body such as iodine, iron, calcium and phosphorus [2]. In addition, its medicinal value is significant because it contains numerous bioactive constituents such as polysaccharides, allantoin, and polyphenols including flavonoids [3]. Previous studies have demonstrated that the Chinese yam exerts strong anti-oxidative [4], cholesterol-lowering [5], growth hormone-releasing activity [6], and protective effects against ethanol-induced gastric ulcers [7]. Recent studies have revealed intimate relations between bioactivities and their responsible secondary metabolites including allantoin, saponins and phenanthrenes. Allantoin has been reported to promote wound healing, the speeding up of cell regeneration, and the exhibition of a keratolytic effect [8]. Saponins from the Chinese yam are reported to have anticancer and fungistatic activity [9]. Phenanthrenes are a representative class of phenolic phytochemicals found in the Dioscoreaceae family. They are known to be biosynthesized from the oxidative coupling of the aromatic rings of stilbene precursors [10] by the tubers, roots, and stems of only 17 taxonomical families of plants [11]. Phenanthrene-containing plants including the *Dioscorea* genus have been widely used for the treatment of several diseases in Africa, Asia, and South America [11]. Since the discovery of antifungal phenanthrenes from *D. rotundata* [12], anti-inflammatory [6,13,14], anticholinesterase [15], and triglyceride accumulation inhibitory [16] activity have been reported as medicinal or health-promoting effects of phenanthrenes discovered from the *Dioscorea* genus. In addition to their bioactivity, phenanthrenes have been suggested to be a non-polar standard marker for the *Dioscorea* genus, as their structures are distinguished from common phytochemicals and they are not difficult to analyze with the conventional high performance liquid chromatography with ultraviolet detection (HPLC-UV) method compared to other phytochemicals found in the *Dioscorea* genus such as allantoin and saponins [17].

Our previous studies on biological activities of *D. batatas* also revealed anti-inflammatory [18] and inhibitory effects on the particulate matter-induced pulmonary injury of phenanthrenes [19]. On the basis of cumulative research, phenanthrenes are putative medicinal components in *D. batatas*. However, how much the contents of phenanthrenes are in actual food products provided for consumers of *D. batatas* has been overlooked. The most common postharvest processing of *D. batatas* tuber is peeling, drying under heat, and then pulverizing the dried flesh to make 'yam flour,' which yields discarded peels without utilization [20]; or *D. batatas* is deteriorated so quickly that only a small fraction of harvested *D. batatas* is supplied as raw products, and most of the Chinese yam is consumed as 'yam flour' or products from the flour [21]. The aim of this study was to verify if phenanthrenes, the polyphenolic bioactive components, are properly provided to consumers of the Chinese yam by investigating the contents of three representative phenanthrenes in the *D. batatas* flesh (DBF), an edible portion, and the *D. batatas* peel (DBP), a discarded byproduct.

2. Materials and Methods

2.1. Plant Material

The Chinese yam (*D. batatas*) was purchased from Taesan-nongjang (Andong, Korea). The Chinese yam tubers were peeled off, and then the flesh and the peel were separated. The peel was washed with water. Then, both the flesh and the peel were cut into slices and dried with a freeze-dryer (Ilshinbiobase, Dongducheon, Korea).

2.2. Chemicals and Reagents

Ethyl acetate (EA, extra pure grade), dichloromethane (DCM, extra pure grade), methanol (MeOH, extra pure grade), hexanes (extra pure grade) and butanol (extra pure grade) were purchased from Duksan pure chemicals Co. (Ansan, Korea). Acetonitrile (reagent grade), water (reagent grade) and methanol (reagent grade) were purchased from J.T.Baker (Phillipsburg, NJ, USA). Acetone-d6 (deuteration degree min. 99.9%) and dimethyl sulfoxide- d_6 (deuteration degree min. 99.8%) for NMR spectroscopy were purchased from Merck (Darmstadt, Germany). Trifluoroacetic acid (TFA), 1,1-diphenyl-2-picryl hydrazyl (DPPH), 2,2-azino-bis-(3-ethylbenzothiazoline-6-sulfonic acid), and diammonium salt (ABTS+) were purchased from Sigma-Aldrich (St Louis, MO, USA).

2.3. Preparation of Standard Phenanthrene Compounds

Phenanthrene Compounds **1**, **2** and **3** (Figure 1) were isolated from the peel of *D. batatas* following the method suggested by our previous studies [18,19]. The peel was extracted with 95% ethanol for 48 h, and solvents were removed in vacuo. The ethanol extract was partitioned into n-hexane, ethyl acetate, butanol and water layers. The ethyl acetate layer was dissolved in mixture of DCM:MeOH (1:1, *v/v*) and fractioned on a size exclusion chromatography column filled with Sephadex® LH20 resin (Pharmacia, Stockholm, Sweden) using DCM:MeOH (1:1, *v/v*) as a mobile phase. Among the thirteen fractions yielded, the eighth fraction was further separated with a high performance liquid chromatograph (Waters 1525 system, Waters, Milford, MA, USA) equipped with a dual absorbance ultraviolet detector (Waters 2487, Waters). A gradient mixture of acetonitrile/water (39:61 (0–44 min) → 100:0 (44.01–60 min)) was used as a mobile phase, and a Hector-A-C18 column (250 × 10.0 mm, 5 μm, RStech corporation, Daejeon, Korea) was used to isolate Compounds **1** (2,7-dihydroxy-4,6-dimethoxy phenanthrene), **2** (6,7-dihydroxy-2,4-dimethoxy phenanthrene). Compound **3** was isolated from the ninth fraction from LH20 column chromatography by using the same HPLC condition for the isolation of Compounds **1** and **2**. The structures of Compounds **1** and **2** were identified by our previous study [18,19] on the basis of ^{1}H- and ^{13}C-NMR spectra (Figures S1 and S2), and the structure of Compound **3** as identified as 6-hydroxy-2,4,7-trimethoxyphenanthrene (batatasin I) through a comparison of ^{1}H and ^{13}C NMR spectra (Figure S3) with a previous report [22]. NMR experiments were carried out using a Bruker Ascend (^{1}H-500 MHz, ^{13}C-125 MHz, Billerica, MA, USA) spectrometer.

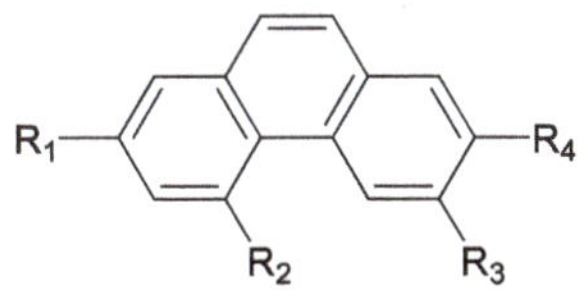

	R$_1$	R$_2$	R$_3$	R$_4$
1	OH	OCH$_3$	OCH$_3$	OH
2	OCH$_3$	OCH$_3$	OH	OH
3	OCH$_3$	OCH$_3$	OH	OCH$_3$

Figure 1. Structures of Phenanthrenes **1–3**.

2.4. Samples and Standard Solutions Preparation

The freeze-dried *D. batatas* flesh (DBF) and *D. batatas* peel (DBP) were powdered, and 1 g of dried powder from each sample was extracted at 25 °C with 250 mL of 95% ethanol for 12 h, followed by filtration with 0.45 μm syringe filter (Advantec, Tokyo, Japan) and evaporation in vacuo. The resultant extracts were weighed and dissolved in 10 mL of methanol. The stock solutions of Compounds **1–3** were prepared by dissolving Compounds **1–3** in methanol to yield 1 mg/mL. Standard solutions were prepared by the serial dilution of the stock solutions to methanol. The range of the standard solutions was 3.125–100 μg/mL (3.125, 6.25, 12.5, 25, 50 and 100 μg/mL). All of the standard solutions were stored at −20 °C in darkness until analysis.

2.5. Quantitative Analysis of Phenanthrenes with HPLC

The quantitative analysis of Phenanthrene Compounds **1–3** was performed using an Alliance 2695 HPLC system (Waters) with a photodiode array detector (Waters 2996, Waters) and a Hector-M-C18 column (250 × 4.6 mm, 5 μm, RStech). The overall method was modified from the method using HPLC-UV suggested by Yoon et al. [17]. The mobile phases used were 0.1% trifluoroacetic acid in water (A) and acetonitrile (B) at a flow rate of 1.0 mL/min. The gradient program was set as follows: 0–5 min, 5% B; 5–20 min, 5%–100% B; and 20–25 min, 100% B. The column temperature was maintained at 40 °C throughout the analysis. The chromatograms were monitored at a wavelength of 260 nm, and the injection volume was 10 μL. The DBF and DBP samples were injected 3 times each, and the averages of the peak areas on the chromatograms were obtained for quantitative analysis. Calibration curves of each compounds were obtained by averaging the peak areas of each compound on the chromatograms acquired from 3 injections that were monitored at a wavelength of 260 nm. The limit of detection (LOD) and the limit of quantification (LOQ) under the chromatographic conditions were determined by the serial dilution of the standard solution on the basis of a signal to noise (S/N) ratio of 3 to 10.

2.6. Determination of DPPH Radical Scavenging Activity

The DPPH radical scavenging activity of the extracts and compounds were measured as described by Kirigaya et al. [23] with slight modifications. Twenty microliters of each extract were mixed with 80 µL of the 0.2 mM DPPH radical methanolic solution. After incubation at 37 °C for 30 min in darkness, the absorbance of the mixture was measured at 515 nm using a Sunrise™ microplate reader (Tecan Group Ltd., Männedorf, Switzerland). To evaluate the radical scavenging activity of Compounds **1–3**, each compound was dissolved in 70%ethanol (*v/v*) at concentrations of 0.0125–1.000 mg/mL. For extracts, DBP and DBF, they were dissolved in 70% ethanol (*v/v*) at concentrations of 0.125–10.000 mg/mL. *L*-ascorbic acid in 70% ethanol (*v/v*) was used as a positive control at concentrations of 0.0125–0.1000 mg/mL. The effective concentrations of the compounds or extracts required to scavenge DPPH radical by 50% (IC_{50}) were obtained by a linear regression analysis of the dose–response curve plotting between %inhibition and concentrations. The DPPH radical scavenging activity at every concentration of each compound or extract was documented in Table S1.

2.7. ABTS+ Radical Scavenging Activity

The radical cations were prepared by mixing a 7 mM ABTS+ stock solution in water with 2.45 mM potassium persulfate and stored in the dark for 12–16 h at room temperature. The ABTS+ solution was diluted with absolute ethanol to an absorbance of 0.7 ± 0.02 at 734 nm before use. In a 96-well plate, 20 µL of each extract was mixed with an 80 µL ABTS+ solution. Measurements were taken at 734 nm using a Sunrise™ microplate reader (Tecan Group Ltd.). To evaluate the radical scavenging activity of Compounds **1–3**, each compound was dissolved in 70% ethanol (*v/v*) at concentrations of 0.0125–1.000 mg/mL. *L*-ascorbic acid in 70% ethanol (*v/v*) was used as a positive control at concentrations of 0.0125–0.1000 mg/mL. The ABTS+ radical scavenging activity of the DBP and DBF extracts dissolved in 70% ethanol (*v/v*) at concentrations of 0.125–10.000 mg/mL. The effective concentrations of the compounds or extracts required to scavenge ABTS+ radical by 50% (IC_{50}) were obtained by a linear regression analysis of the dose–response curve plotting between %inhibition and concentrations. The ABTS+ radical scavenging activity at every concentration of each compound or extract was documented in Table S2.

2.8. Statistical Analysis

The experimental results are presented as mean and standard deviation (mean ± SD). All the experiments were analyzed in triplicated measurements. The one-way analysis of variance (ANOVA) was run using the SPSS ver. 23.0 software (SPSS Inc., Chicago, IL, USA), and the Duncan's multiple range test comparisons at $p < 0.05$ were run to determine significant difference.

3. Results and Discussion

3.1. Calibration, Limit of Detection (LOD) and Limit of Quantification (LOQ)

The method used in this study was an HPLC-UV method that monitored phenanthrene derivatives at the wavelength of 260 nm. A similar method was suggested and validated by Yun et al. [17]. In this study, linearity, the LOD and the LOQ were validated. For the preparation of the calibration curve, standard solutions were prepared by serial dilution to appropriate concentrations. The linearity of calibration curves obtained from standard solutions was satisfactory with the determination coefficients (R^2), which were greater than 0.9959 in the concentration range of 3.125–100 µg/mL (Table 1). The

LODs of standard phenanthrenes were smaller than 0.58 µg/mL, and the LOQs were smaller than 1.94 µg/mL (Table 1).

Table 1. Calibration parameters for the HPLC method.

Compound	Regression Equation	Range (µg/mL)	R^2	LOD (µg/mL)	LOQ (µg/mL)
1	Y = 301.34x − 46,588	3.125–100	0.9994	0.44	1.47
2	Y = 510.71x − 21,040	3.125–100	0.9990	0.58	1.94
3	Y = 99.155x + 15,3981	3.125–100	0.9959	0.42	1.40

3.2. Quantitative Analysis of Phenanthrenes with HPLC

Chemical screening for phenanthrenes present in the DBP and DBF extracts by monitoring the HPLC chromatograms of 260 nm wavelength detection revealed that Compounds **1–3** are major phenanthrenes present in the DBP, but the DBF scarcely contained any type of phenanthrenes (Figure 2). The chromatogram of the DBF extract exhibited a peak at the retention time (r.t.) of 12.10 min, and its UV absorption spectrum was completely different from those of phenanthrenes (data not shown). According to the chromatograms, Compounds **1–3** were completely separated without overlapping with other peaks. The retention times (r.t.) of Compounds **1–3** were 16.4, 17.5 and 19.2 min, respectively.

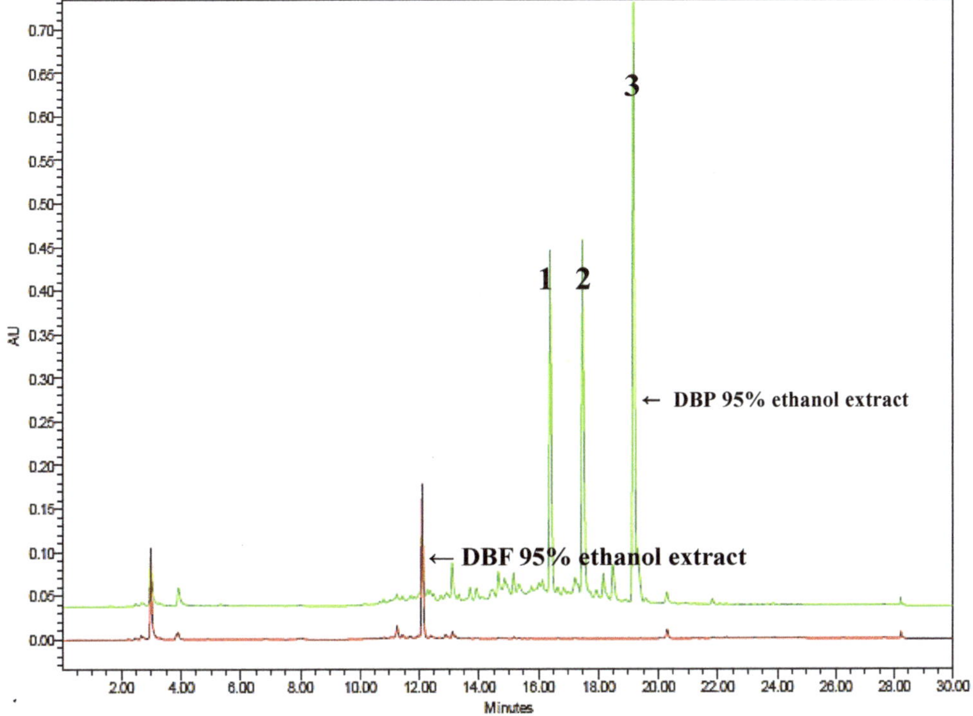

Figure 2. HPLC chromatogram of the *Dioscorea batatas* flesh (DBF) (red) and the *D. batatas* peel (DBP) (green) extracts monitored at the wavelength of 260 nm.

By the subsequent quantitative analysis with HPLC, the content of Compound **1** in the DBP was 47.35 ± 0.25 mg/100 g on the dry weight basis and 6.76 ± 0.04 mg/100 g on the wet weight basis; the content of Compound **2** was 29.29 ± 0.08 mg/100 g on the dry wt. basis and 4.18 ± 0.01 mg/100 g on the wet wt. basis; the content of Compound **3** was 35.85 ± 0.12 mg/100 g on the dry wt. basis and 5.12 ± 0.02 mg/100 g on the wet wt. basis. However, it was not possible to assess the contents of any phenanthrenes in the DBF extract, as the DBF contained only trace amounts under the detection limit values of Phenanthrenes **1–3** (Table 2). There already have been attempts to investigate differences in antioxidant activity and total phenolic contents between the flesh and the peel of the Chinese yam [24,25]. Both of these research articles reported that the Chinese yam peel exhibited more potent antioxidant activity and more total phenolic contents than the flesh. The study of Liu et al. [25] also revealed a significant difference between total phenolics and total flavonoids contents in the peel, which implies that the peel contains even more amounts of non-flavonoidal phenolics than flavonoids, at least by five folds. Their study did not identify the responsible non-flavonoidal phenolic compounds that account for the difference. However, from our analytical result, it is now comprehensible to presume that phenanthrenes may largely contribute to the phenolics contents in the Chinese yam peel.

Table 2. Contents of Phenanthrenes **1–3** in the DBF and DBP 95% ethanol extracts.

		Content (mg/100 g)			
		DBP		DBF	
		Dry wt. Basis	Wet wt. Basis	Dry wt. Basis	Wet wt. Basis
	1	47.35 ± 0.25	6.76 ± 0.04	N.D.	N.D.
Compound	2	29.29 ± 0.08	4.18 ± 0.01	N.D.	N.D.
	3	35.85 ± 0.12	5.12 ± 0.02	N.D.	N.D.

* Values represent the means ± SD ($n = 3$).

3.3. Antioxidant Activity of Phenanthrenes, DBP and DBF

As presented in Figure 3, Compound **1** exhibited the strongest antioxidant activity among the phenanthrenes isolated from *D. batatas*. Compound **1** exhibited strong DPPH radical scavenging activity with an IC_{50} value of 0.0645 mg/mL and an even stronger ABTS+ inhibition with an IC_{50} value of 0.0482 mg/mL, which was equivalent to the ABTS+ inhibition of the positive control, *L*-ascorbic acid (IC_{50} 0.499 mg/mL). Compounds **2** and **3** also exhibited relatively weaker antioxidant activity compared to Compound **1** and *L*-ascorbic acid. Compound **2** exhibited DPPH and ABTS+ radical scavenging activity with IC_{50} values of 0.154 and 0.153 mg/mL, respectively. Compound **3** exhibited DPPH and ABTS+ radical scavenging activity with IC_{50} values of 0.566 and 0.297 mg/mL, respectively. The strong antioxidant activity of Compound **1** is not surprising, as it has been reported to be an antioxidant [14], an antifungal [26] and a pancreatic lipase inhibitory agent [27] isolated from plants that belong to *Dioscorea* sp. cultivated in Korea. Our previous study on the anti-inflammatory activity of *D. batatas* [18] also led to the isolation of Compound **1** as a responsible component for the bioactivity.

In accordance with previous studies that have revealed the difference of antioxidant activity between the peel and the flesh of the Chinese yam [24,25], the DPPH and ABTS+ scavenging assay exhibited far more potent antioxidant activity of the DBP compared to that of the DBF (Figure 3). Both the DPPH and ABTS+ radical scavenging assays showed a wide gap in antioxidant activity between the DBP and the DBF (Figure 3). The DBP extract dissolved in 70% ethanol exhibited inhibition against DPPH and ABTS+ with IC_{50} values of 0.944 and 0.771 mg/mL, respectively. On the other hand, the DBF exhibited far weaker antioxidant activity, showing inhibition against DPPH with an IC_{50} value of 7.68 mg/mL and inhibition against ABTS+ with an IC_{50} value of 3.43 mg/mL. A previous report by Liu et al. [25] doubtfully concluded that the more potent antioxidant activity of the peel might be attributed to the slightly higher content of allantoin in the peel than in the flesh. However, the results from this study imply that the difference in the content of phenanthrenes, antioxidant agents as potent as the positive control *L*-ascorbic acid, is the crucial factor for the large difference in the antioxidant activity between the DBP and the DBF.

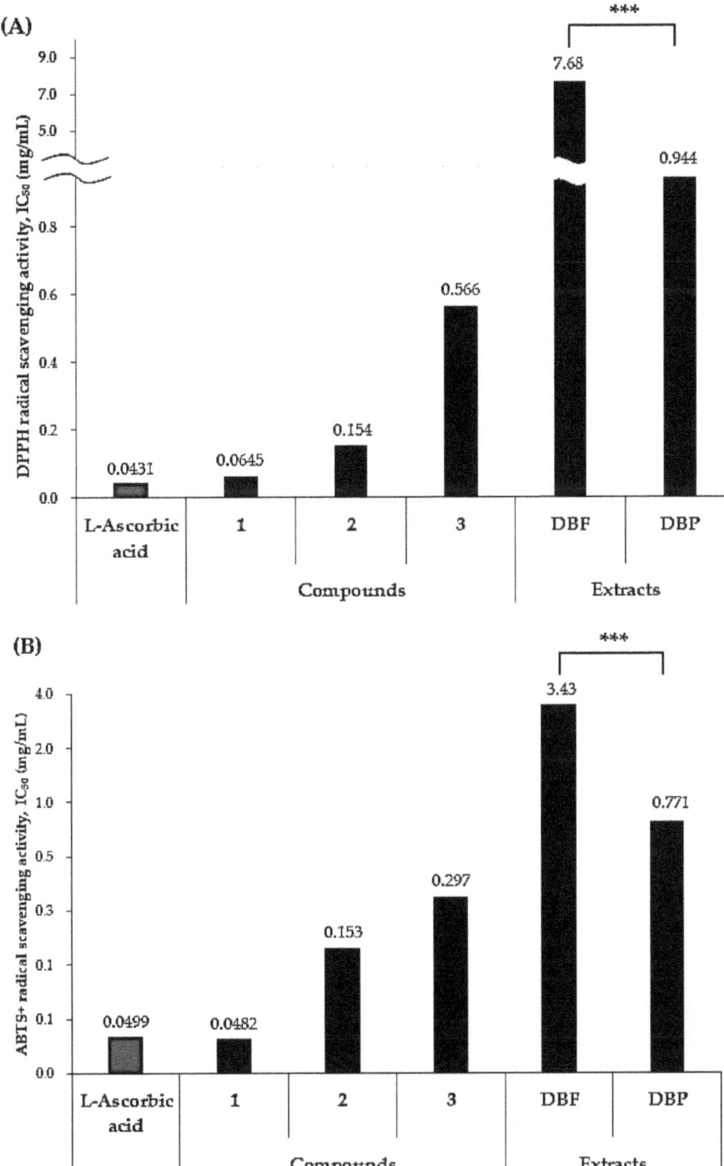

Figure 3. Antioxidant activities of Compounds **1–3** for the DBP and DBF extracts. The DPPH (diphenyl-2-picryl hydrazyl) (**A**) and ABTS+ (diammonium salt) (**B**) radical scavenging activities for Compounds **1–3**, the DBP, the DBF extracts, and the positive control (L-ascorbic acid) were measured in IC$_{50}$ (mg/mL). *** $p < 0.001$.

4. Conclusions

In practice, the Chinese yam peel has been regarded as inedible due to its soil contents, and it has been discarded without utilization. As the Chinese yam has a low preservation potential, most of its harvest biomass is to be peeled, dried, and powdered. The analysis of phenanthrene contents in the DBP compared to the DBF reflects that consumers hardly ingest potent antioxidant agents suchas phenanthrenes by consuming products from the Chinese yam flour made by conventional postharvest processing. However, the Chinese yam peel has now been proven a valuable resource for bioactive components with high potentials for pharmaceuticals or functional biomaterials. In conclusion, a quantitative analysis of phenanthrenes, representative phenolic natural products with high antioxidant and functional potentials, was performed using the HPLC-UV method, which revealed that the contents in the DBP are significantly higher than in the DBF. The difference in the contents of phenanthrenes is estimated to result in the considerable difference of antioxidant potential between the DBP and the DBF. Further to the results from this study, it is strongly urged to investigate the appropriate means to optimize the utilization of the discarded biomass of the Chinese yam peel.

Supplementary Materials: The following are available online at http://www.mdpi.com/2076-3921/8/11/541/s1, Figure S1: ^{1}H and ^{13}C NMR spectrum of Compound **1**, Figure S2: ^{1}H and ^{13}C NMR spectrum of Compound **2**, Figure S3: ^{1}H and ^{13}C NMR spectrum of Compound **3**, Table S1: DPPH radical scavenging activity of Compounds **1–3**, the DBP and DBF extracts., Table S2: ABTS+ radical scavenging activity of Compounds **1–3**, the DBP and DBF extracts.

Author Contributions: Conceptualization, J.-G.L., D.H. and J.-S.B.; methodology, M.J.G. and J.-G.L.; validation, M.K. and J.-G.L.; formal analysis, M.K. and M.J.G.; investigation, M.K. and M.J.G.; resources, M.J.G.; software, J.C. and M.J.G.; data curation, M.K., J.C., and J.-S.B.; writing—original draft preparation, M.K., M.J.G. and D.H.; writing—review and editing, D.H., J.C and J.-S.B; visualization, M.K., M.J.G., J.-G.L. and J.-S.B; supervision, D.H. and J.-S.B.; funding acquisition, D.H.

Funding: This research was supported by the National Research Foundation of Korea (NRF) grant funded by the Korean government (2019R1F1A1051041).

Acknowledgments: The authors acknowledge Forest Resources Development Institute of Gyeongsangbuk-do (Andong, Korea) for advice on research.

Conflicts of Interest: The authors declare no conflict of interest.

References

1. Ryu, H.-Y.; Bae, K.-H.; Kim, E.-J.; Park, S.-J.; Lee, B.-H.; Sohn, H.-Y. Evaluation for the Antimicrobial, Antioxidant and Antithrombosis Activity of Natural Spices for Fresh-cut yam. *J. Life Sci.* **2007**, *17*, 652–657. [CrossRef]

2. Sukhija, S.; Singh, S.; Riar, C.S. Effect of oxidation, cross-linking and dual modification on physicochemical, crystallinity, morphological, pasting and thermal characteristics of elephant foot yam (*Amorphophallus paeoniifolius*) starch. *Food Hydrocolloid* **2016**, *55*, 56–64. [CrossRef]

3. Yang, W.F.; Wang, Y.; Li, X.P.; Yu, P. Purification and structural characterization of Chinese yam polysaccharide and its activities. *Carbohyd. Polym.* **2015**, *117*, 1021–1027. [CrossRef] [PubMed]

4. Hou, W.C.; Hsu, F.L.; Lee, M.H. Yam (*Dioscorea batatas*) tuber mucilage exhibited antioxidant activities in vitro. *Planta Med.* **2002**, *68*, 1072–1076. [CrossRef] [PubMed]

5. Oh, M.; Houghton, P.; Whang, W.; Cho, J. Screening of Korean herbal medicines used to improve cognitive function for anti-cholinesterase activity. *Phytomedicine* **2004**, *11*, 544–548. [CrossRef] [PubMed]

6. Lu, Y.; Jin, M.H.; Park, S.J.; Son, K.H.; Son, J.K.; Chang, H.W. Batatasin I, a Naturally Occurring Phenanthrene Derivative, Isolated from Tuberous Roots of *Dioscorea batatas* Suppresses Eicosanoids Generation and Degranulation in Bone Marrow Derived-Mast Cells. *Biol. Pharm. Bull.* **2011**, *34*, 1021–1025. [CrossRef] [PubMed]

7. Byeon, S.; Oh, J.; Lim, J.S.; Lee, J.S.; Kim, J.S. Protective Effects of *Dioscorea batatas* Flesh and Peel Extracts against Ethanol-Induced Gastric Ulcer in Mice. *Nutrients* **2018**, *10*, 1680. [CrossRef]

8. Zhang, Z.; Gao, W.; Wang, R.; Huang, L. Changes in main nutrients and medicinal composition of Chinese yam (*Dioscorea opposita*) tubers during storage. *J. Food Sci. Technol.* **2014**, *51*, 2535–2543. [CrossRef]

9. Zhang, H.; Yu, X.; Li, X.; Pan, X. A study of promotive and fungistatic actions of steroidal saponin by microcalorimetric method. *Thermochim. Acta* **2004**, *416*, 71–74. [CrossRef]

10. Chong, J.L.; Poutaraud, A.; Hugueney, P. Metabolism and roles of stilbenes in plants. *Plant Sci.* **2009**, *177*, 143–155. [CrossRef]

11. Toth, B.; Hohmann, J.; Vasas, A. Phenanthrenes: A Promising Group of Plant Secondary Metabolites. *J. Nat. Prod.* **2018**, *81*, 661–678. [CrossRef] [PubMed]

12. Coxon, D.T.; Ogundana, S.K.; Dennis, C. Antifungal phenanthrenes in yam tubers. *Phytochemistry* **1982**, *21*, 1389–1392. [CrossRef]

13. Li, Q.; Zhang, C.R.; Dissanayake, A.A.; Gao, Q.Y.; Nair, M.G. Phenanthrenes in Chinese Yam Peel Exhibit Antiinflammatory Activity, as Shown by Strong in Vitro Cyclooxygenase Enzyme Inhibition. *Nat. Prod. Commun.* **2016**, *11*, 1313–1316. [CrossRef] [PubMed]

14. Yang, M.H.; Yoon, K.D.; Chin, Y.W.; Park, J.H.; Kim, J. Phenolic compounds with radical scavenging and cyclooxygenase-2 (COX-2) inhibitory activities from *Dioscorea opposita*. *Bioorg. Med. Chem.* **2009**, *17*, 2689–2694. [CrossRef] [PubMed]

15. Boudjada, A.; Touil, A.; Bensouici, C.; Bendif, H.; Rhouati, S. Phenanthrene and dihydrophenanthrene derivatives from *Dioscorea communis* with anticholinesterase, and antioxidant activities. *Nat. Prod. Res.* **2018**, 1–5. [CrossRef]

16. Du, D.; Zhang, R.; Xing, Z.; Liang, Y.; Li, S.; Jin, T.; Xia, Q.; Long, D.; Xin, G.; Wang, G. 9, 10-Dihydrophenanthrene derivatives and one 1, 4-anthraquinone firstly isolated from *Dioscorea zingiberensis* CH Wright and their biological activities. *Fitoterapia* **2016**, *109*, 20–24. [CrossRef]

17. Yoon, K.D.; Yang, M.H.; Nam, S.I.; Park, J.H.; Kim, Y.C.; Kim, J.W. Phenanthrene derivatives, 3, 5-dimethoxyphenanthrene-2, 7-diol and batatasin-1, as non-polar standard marker compounds for Dioscorea Rhizoma. *Nat. Prod. Sci.* **2007**, *13*, 378–383.

18. Lim, J.S.; Hahn, D.; Gu, M.J.; Oh, J.; Lee, J.S.; Kim, J.-S. Anti-inflammatory and antioxidant effects of 2, 7-dihydroxy-4, 6-dimethoxy phenanthrene isolated from *Dioscorea batatas* Decne. *Appl. Biol. Chem.* **2019**, *62*, 29. [CrossRef]

19. Lee, W.; Jeong, S.Y.; Gu, M.J.; Lim, J.S.; Park, E.K.; Baek, M.-C.; Kim, J.-S.; Hahn, D.; Bae, J.-S. Inhibitory effects of compounds isolated from *Dioscorea batatas* Decne peel on particulate matter-induced pulmonary injury in mice. *J. Toxicol. Env. Heal. A* **2019**, 1–14. [CrossRef]

20. Kwon, J.-B.; Kim, M.-S.; Sohn, H.-Y. Evaluation of antimicrobial, antioxidant, and antithrombin activities of the rhizome of various *Dioscorea* species. *Korean J. Food Preserv.* **2010**, *17*, 391–397.

21. Lee, H.; Seo, D.-H.; Kim, H.-S. Effect of starch extraction solutions on extraction and physicochemical property of Chinese yam (*Dioscorea batatas*) starch. *Korean J. Food Sci. Techol.* **2018**, *50*, 191–197. [CrossRef]

22. Kum, E.-J.; Park, S.-J.; Lee, B.-H.; Kim, J.-S.; Son, K.-H.; Sohn, H.-Y. Antifungal activity of phenanthrene derivatives from aerial bulbils of *Dioscorea batatas* Decne. *J. Life Sci.* **2006**, *16*, 647–652. [CrossRef]

23. Kirigaya, N.; Kato, H.; Fujimaki, M. Studies on antioxidant activity of nonenzymic browning reaction products. Part 3. Fractionation of browning reaction solution between ammonia and D-glucose and antioxidant activity of the resulting fractions. *Nippon Nogeikagaku Kaishi* **1971**, *45*, 292–298. [CrossRef]

24. Chung, Y.C.; Chiang, B.H.; Wei, J.H.; Wang, C.K.; Chen, P.C.; Hsu, C.K. Effects of blanching, drying and extraction processes on the antioxidant activity of yam (*Dioscorea alata*). *Int. J. Food Sci. Tech.* **2008**, *43*, 859–864. [CrossRef]

25. Liu, Y.; Li, H.; Fan, Y.; Man, S.; Liu, Z.; Gao, W.; Wang, T. Antioxidant and Antitumor Activities of the Extracts from Chinese Yam (*Dioscorea opposite* Thunb.) Flesh and Peel and the Effective Compounds. *J. Food Sci.* **2016**, *81*, H1553–H1564. [CrossRef] [PubMed]

26. Takasugi, M.; Kawashima, S.; Monde, K.; Katsui, N.; Masamune, T.; Shirata, A. Antifungal compounds from *Dioscorea batatas* inoculated with *Pseudomonas cichorii*. *Phytochemistry* **1987**, *26*, 371–375. [CrossRef]

27. Yang, M.H.; Chin, Y.-W.; Yoon, K.D.; Kim, J. Phenolic compounds with pancreatic lipase inhibitory activity from Korean yam (*Dioscorea opposita*). *J. Enzyme Inhib. Med. Chem.* **2014**, *29*, 1–6. [CrossRef]

 *antioxidants*

Article

Antiplatelet Activity of Natural Bioactive Extracts from Mango (*Mangifera Indica* L.) and its By-Products

María Elena Alañón [1,2,3,]*, Iván Palomo [4,]*, Lyanne Rodríguez [4], Eduardo Fuentes [4], David Arráez-Román [2,3,†] and Antonio Segura-Carretero [2,3,†]

[1] Area of Food Technology, Regional Institute for Applied Scientific Research (IRICA), University of Castilla-La Mancha. Avda. Camilo José Cela, 10, 13071 Ciudad Real, Spain

[2] Department of Analytical Chemistry, Faculty of Sciences, University of Granada, C/Fuentenueva s/n, 18071 Granada, Spain; darraez@ugr.es (D.A.-R.); ansegura@ugr.es (A.S.-C.)

[3] Research and Development of Functional Food Centre (CIDAF), PTS Granada, Avda. Del Conocimiento 37, Bioregión Building, 18016 Granada, Spain

[4] Thrombosis Research Center, Department of Clinical Biochemistry and Immunohaematology, Faculty of Health Sciences, Interdisciplinary Center on Aging (CIE), University of Talca, 3460000 Talca, Chile; Lyannerodriguez89@gmail.com (L.R.); efuentes@utalca.cl (E.F.)

* Correspondence: mariaelena.alanon@uclm.es (M.E.A.); ipalomo@utalca.cl (I.P.)

† Both authors are joint senior researchers on this work.

Received: 2 October 2019; Accepted: 26 October 2019; Published: 29 October 2019

Abstract: The potential antiplatelet aggregation effects of mango pulp and its by-products (peel, husk seed, and seed) due to the presence of bioactive compounds were explored. Among them, mango seed exhibited a 72% percentage inhibition of platelet aggregation induced by adenosine 5′-diphosphate (ADP) agonist with a demonstrated dose-dependent effect. This biological feature could be caused by the chemical differences in phenolic composition. Mango seed was especially rich in monogalloyl compounds, tetra- and penta-galloylglucose, ellagic acid, mangiferin, and benzophenones such as maclurin derivatives and iriflophenone glucoside. Mangiferin showed an inhibitory effect of 31%, suggesting its key role as one of the main contributors to the antiplatelet activity of mango seed. Therefore, mango seed could be postulated as a natural source of bioactive compounds with antiplatelet properties to design functional foods or complementary therapeutic treatments.

Keywords: Mango; by-products; antiplatelet activity; bioactive compounds; HPLC-DAD-q-TOF-MS

1. Introduction

Cardiovascular diseases (CVD) have the highest mortality rate of all types of diseases worldwide. Atherosclerosis and thrombotic processes associated with the rupture of vulnerable plaques are the main triggers of cardiovascular and cerebrovascular strokes [1,2]. It has been determined that platelets represent the bridge between inflammation and thrombosis, which are fundamental processes in the development of atherothrombosis [3]. The general pathogenies entails platelet activation, subsequent adhesion, release of granule content, and platelet aggregation [4]. The multiple mechanisms of action and the side effects of drugs make the research on natural bioactive compounds useful for pharmaceutical constituents in the prevention or complementary treatment of antiplatelet therapy.

The antiplatelet activity of numerous bioactive compounds detected in fruits and vegetables and their multiple mechanisms of actions have recently been highlighted [5]. However, these bioactive compounds can not only be found in the edible fraction but also in their by-products. This fact has made by-products a profitable niche of phytochemicals to be used as functional substances [6], although some of their characteristics such as bioavailability, pharmakinetic, pharmacodynamics properties, safety, and toxicity still remain unclear.

In this context, mango fruit (*Mangifera indica* L) is the crop with the second highest production and acreage requirements, behind only bananas [7]. Although it is mainly consumed fresh, plenty of mango-derived products have been created such as juice, nectar, purée, ice cream, jam, canned slices, chutneys, etc. Consequently, by-products such as peel, seed, and seed husk, constituting 35%–60% of the fruit [8], are generated during the industrial processes. Recently, the high content of health-enhancing compounds like polyphenols, anthocyanins, and carotenoids in mango by-products has been reported [9]. Thus, different scientific investigations have been focused on the revalorization of mango by-products due to their functional properties and potential therapeutic uses of their bioactive compounds [10,11].

Recently, the antioxidant and anti-inflammatory effects of phenolic compounds obtained from mango by-products and water extracts by reducing the nitric oxide (NO) levels produced by lipopolysaccharides (LPS)-stimulated macrophages have been reported [12]. This biological activity postulates mango by-products to enhance cardiovascular health, and makes other physiological effects of mango by-products plausible against the pathogenies of cardiovascular diseases.

Therefore, the main goal of this work is to explore the inhibitory effects of platelet aggregation of extracts from mango pulp and its by-products (peel, husk seed, and seed) and characterize the phenolic composition in order to figure out the compounds responsible for the antiplatelet activity.

2. Materials and Methods

2.1. Fruit Material and Sample Preparation

About 20 kg of Keitt mangos cultivated in the Tropical Coast of Granada were provided by Miguel García Sanchez e Hijos S.A. (Motril, Spain) at the optimum maturation stage (13.1–16.0 °Brix) in November 2016. Once they had arrived at the laboratory, samples were cleaned and the different parts of mango (pulp, peel, seed, and seed husk) were separated manually and cut into small and homogenous pieces. To preserve chemical composition of mango pulp and its by-products, samples were submitted to a lyophilization process (Advantage Plus EL-85 freeze dryer, SP Scientific, Ipswich, Suffolk, UK). Then each sample was milled (IKA M20-IKAWERKE GmbH & Co. KG, Staufen, Germany), homogenized, and stored at −18 °C prior to their analyses.

2.2. Isolation of Extracts Rich in Polyphenols from Mango and its By-Produts

2.2.1. Extracts for Anti-Platelet Aggregation Activity Assay

The isolation of polyphenols from samples was carried out following an adapted method from Gómez-Caravaca et al., 2016 [13]. For each sample, 2 g of freeze-dried powder were sonicated for 15 min with 40 mL of a solution of methanol/water (80:20 v/v). After the extraction process, the mixture was centrifuged for 15 min at 7700 g and at 4 °C. The supernatant was removed and other two consecutive extraction steps were repeated. All supernatants were collected, evaporated to dryness, and stored at −18 °C. With the aim of obtaining arond 2 g of each extract, we performed the antiplatelet activity and took into account the yield of each extraction (pulp, 79.4%; peel, 44.6%; seed husk, 7.4%, and husk, 16.7%). In total, 4 g of pulp, 6 g of peel, 32 g of seed husk, and 16 g of husk were processed using the previous extraction procedure.

2.2.2. Extracts for Analytical Characterization

For each samples (pulp, peel, seed husk, and seed) 0.5 g of freeze-dried powder were extracted with 10 mL of methanol/water (80:20 v/v) under sonication for 15 min [13]. After the extraction process, mixture was centrifuged for 15 min at 7700 g and at 4 °C. The supernatant was removed and the extraction step was then repeated twice. Finally, all supernatants were also collected, evaporated, reconstituted in 3 mL of methanol/water (80:20 v/v), filtered (0.2 μM, Milllipore), and stored at −18 °C. All extractions were performed in duplicate.

2.3. Human Platelet Isolation

Platelet-rich plasma (PRP) was obtained from six healthy young volunteers who had previously signed a consent report. Samples were extracted by phlebotomy in 3.2% sodium citrate tubes (Becton Dickinson Vacutainer Systems, Franklin Lakes, NJ, USA). Then, blood samples were centrifuged (DCS-16 Centrifugal Presvac RV) at 240 g for 10 min to obtain platelet-rich plasma (PRP). Two-thirds of PRP was removed and centrifuged at 650 g for 10 min. The platelet count was performed in a blood count (Bayer Advia 60 Hematology System, Tarrytown, NY, USA).

2.4. Anti-Platelet Aggregation Activity Assay

The anti-platelet aggregation activity of mango and its by-products was evaluated by a turbidimetric method [14] using a lumi-aggregometer (Chrono-Log, Haverton, PA, USA). The PRP (200×10^9 platelets/L) was preincubated with 20 µL of phosphate-buffered saline, PBS, (negative control: maximum aggregation), or power extracts from each part of the mango (1 mg/mL) for 3 min at 37 °C. In order to evaluate the dose-dependent effects against platelet aggregation, different concentrations (0.1, 0.5, and 1 mg/mL) of the more active extract were tested as well as the antiplatelet properties of some standards compounds such as mangiferin (Sigma-Aldrich, St. Louis, Missouri, MO, USA) at 1 mg/mL. For all assays, platelet aggregation was induced by adding adenosine 5′-diphosphate (ADP, 4 µM) supplied by Sigma-Aldrich (St. Louis, Missouri, MO, USA) as an agonist. Platelet aggregation was measured by triplicate as the increase in light transmission occurred for 6 min and results were expressed as a percentage of inhibition of aggregation.

2.5. Phenolic Characterization of Extracts by HPLC-DAD-q-TOF-MS

Power extracts from the extraction of 0.5 g of samples were redissolved in 3 mL of methanol/water (80:20 v/v). After their filtration, extracts were analyzed by HPLC-DAD-q-TOF-MS (Agilent 1200 series coupled to 6540 Agilent Ultra-High-Definition Accurate-Mass q-TOF-MS, Agilent Technologies, Palo Alto, CA, USA) according to the method proposed by Gómez-Caravaca et al. [13].

Identification was performed based on relative retention times, UV-Vis spectra, and mass spectra obtained by q-TOF-MS and from the literature. Quantification was carried out using calibration standards curves of gallic acid, coumaric acid, ferulic acid, vanillic acid, catechin, quercetin, ellagic acid, and mangiferin (Sigma Aldrich, St. Louis, MO, USA). Spectral and chromatographic data are compliled in Supplementary Table S1. Integration and data processing were performed using Mass Hunter Workstation Software, Qualitative Analysis, version B.07.00 (Agilent Technologies, Inc. 2014).

2.6. Statistical Analysis

The IBM SPSS statistics v.22.0 for Windows statistical package was employed to carry out statistical analysis. Anti-platelet aggregation data set was submitted to the ANOVA analysis and the Tukey's post-hoc test, while the Student-Newman-Keuls test was applied to the chemical data in order to determine the significant differences between samples.

3. Results and Discussion

3.1. Anti-Platelet Aggregation Activity of Mango and its By-Products

The anti-platelet aggregation activity of extracts from edible part of mango (pulp) and its by-products (peel, seed husk, and seed) was evaluated on a human platelet whose aggregation was induced by using ADP (4 µM) as an agonist. Inhibition percentages of platelet aggregation of extracts from mango and its by-products under the optimum maturation stage at a concentration of 1.0 mg/mL are shown in Figure 1. It can be observed that human platelet aggregation triggered by the ADP agonist was not significantly inhibited by pulp, skin, and husk seed extracts compared to the negative control. However, mango seed extract exhibited a significantly higher inhibition percentage, which was around 72%. These findings can hardly be compared with others due to the scarce bibliography regarding the antiplatelet activity of mango by-products. Recently, the influence of the ripening stage of mango peel on rat platelet aggregation has been reported [15]. Ripe fruit peel extract showed better platelet aggregation inhibitory effects. The inhibition percentage observed in ripened mango peel water extract at 0.8 mg/mL in rat platelets was considerably higher (approximately 25%) in comparison with our results tested against human platelet (approximately 10%). To the best of our knowledge, anti-platelet aggregation data on mango pulp and their by-products on human blood have not been reported until now.

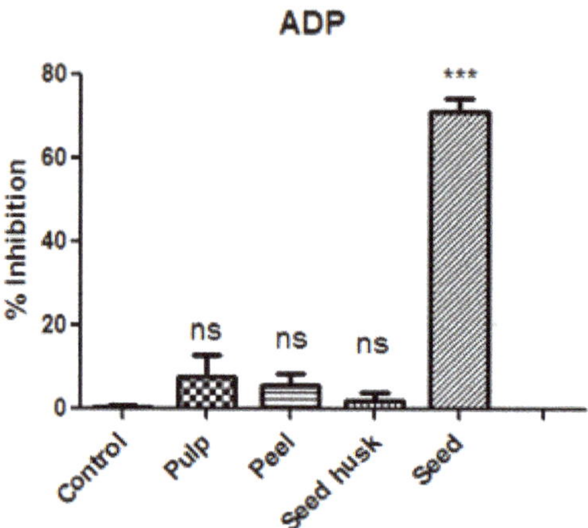

Figure 1. Antiplatelet aggregation activity results of extracts from mango and its by-products (peel, seed husk and seed) at 1.0 mg/mL against adenosine 5′-diphosphate (ADP) agonist (4 µM) expressed as mean value of inhibition percentages (n = 6). Data was analyzed using ANOVA of one factor. Post hoc analyses were conducted using Tukey's test. *** denotes significant differences compared to the negative control (absence of extract) at p = 0.001; ns denotes no statistical differences.

Therefore, based our results, mango seed extract was shown to be a potential inhibitor of human platelet aggregation induced by ADP agonist at 1.0 mg/mL with an inhibition percentage of 72%. The current study also examined the dose-dependent effects of mango seed extract on inhibition of platelet aggregation. To investigate the dose-dependent activity, several concentrations of mango seed extract (1.0, 0.5 and 0.1 mg/mL) were tested against the platelet aggregation induced by ADP agonist (4 µM). Significant differences in antiplatelet aggregation activity were identified between the highest mango seed extract (1.0 mg/mL) and the middle dose (0.5 mg/mL) at p = 0.001 and p = 0.01, respectively (Figure 2). Meanwhile no significant differences were found among the negative control and the lowest dose tested (0.1 mg/mL). Therefore, data showed a clear dose-dependent effect that increased the inhibition of platelet aggregation when the mango seed extract concentration increased.

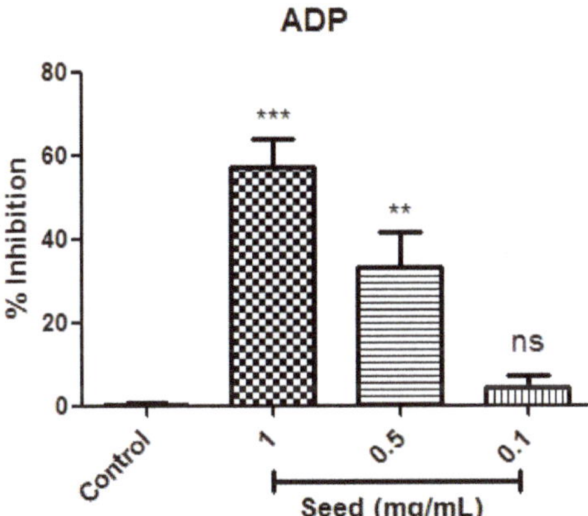

Figure 2. Study of the dose-dependent effect of mango seed extract on antiplatelet aggregation activity against ADP agonist (4 μM) expressed as a mean value of inhibition percentages ($n = 6$). Data were analyzed using ANOVA of one factor. Post hoc analyses were conducted by Tukey's test, ** and *** denotes significant differences at $p = 0.01$ and $p = 0.001$, respectively; ns denotes no statistical differences.

3.2. Phenolic Characterization of Extracts by HPLC-DAD-q-TOF-MS

The nutraceutical and pharmaceutical significance of mango to human health has been attributable to its phenolic composition [11]. Therefore, a phenolic characterization of the edible and non-edible parts of mango was carried out with the aim to relate chemical differences with the antiplatelet activity observed by mango seed. A large number of phenolic compounds with different natures and structures were detected in mango pulp and its by-products. The most abundant and numerous family found in mango was gallic acid derivatives (Table 1). The characteristic high content of gallic acid of mango fruit and its by-products in comparison with other tropical fruits has already been determined previously [12,16].

Mango seed was particularly rich in monogalloyl compounds compared to the rest of mango parts, especially edible mango fraction and seed husk by-product. The sum of the total monogalloyl compounds reached the value of 720.30 ± 2.27 mg/100 g$_{dry\ matter}$ in mango seed. In contrast, considerable lower concentrations of total monogalloyl compounds were found in pulp, peel, and seed husk (52.18 ± 0.10, 235.34 ± 11.07, and 68.15 ± 3.48 mg/100 g$_{dry\ matter}$, respectively). It was noteworthy that the highest content of methylgallate was observed in mango seed (558.86 ± 6.74 mg/100 g$_{dry\ matter}$), which was the major phenolic compound detected in this matrix. The strong antioxidant power of this compound has already been reported in the bibliography [10]. Significantly higher concentrations of other monogalloyl compounds such as gallic acid, galloyl diglucoside, and galloylquinic acid in mango seed were also found. Higher quantities of galloylglucose were found in mango peel, as were monogalloyl derivatives. On the other hand, both edible mango fraction and seed husk exhibited the lowest values of monogalloyl compounds compared to peel and seed by-products. These findings were in good agreement with those reported by other authors, who also detected higher concentrations of gallic acid and methylgallate in mango seed and found more abundant galloylglucose in mango peel [13].

Structurally, gallic acid contains hydroxyl groups and a carboxylic acid group, so its molecules have the ability to react with one another to generate digalloyl compounds. Regarding these compounds, a contrary behavior was evidenced since the highest concentrations of digalloylglucose, methyldigallate, and digalloylquinic acid were found in mango peel (Table 1). Consequently, peel was found to be the fraction of mango with the greatest digalloyl content (137.98 ± 4.09 mg/100 g$_{dry\ matter}$) in comparison with the rest of samples whose composition on digalloyl compounds ranged between 8.13 and 22.50 mg/100 g$_{dry\ matter}$).

Table 1. Mean concentration of gallic acid derivative compounds from different parts of *Keitt* mango variety at optimum maturation stage expressed as mg/100 $g_{dry\ matter}$.

COMPOUNDS	PULP		PEEL		SEED HUSK		SEED	
	MEAN	SD	MEAN	SD	MEAN	SD	MEAN	SD
Monogalloyl compounds								
Gallic acid	1.55	± 0.02 [a]	3.57	± 0.47 [b]	3.93	± 0.24 [b]	16.67	± 0.03 [c]
Galloylglucose	42.36	± 0.23 [a]	112.17	± 1.10 [c]	43.09	± 1.83 [a]	53.67	± 4.73 [b]
Galloyl diglucoside	0.09	± 0.00 [a]	5.69	± 0.07 [b]	0.25	± 0.01 [a]	6.64	± 0.11 [c]
Methylgallate	5.41	± 0.48 [a]	54.14	± 7.77 [b]	14.43	± 0.89 [a]	558.86	± 6.74 [c]
Galloylquinic acid	2.76	± 0.12 [a]	59.77	± 1.79 [c]	6.45	± 0.50 [b]	84.46	± 0.34 [d]
Digalloyl compounds								
Digallic acid	ND		2.54	± 0.34 [a]	1.67	± 0.00 [a]	3.64	± 0.57 [b]
Digalloylglucose	4.59	± 0.03 [a]	31.17	± 0.57 [c]	9.18	± 0.59 [b]	4.23	± 0.09 [a]
Methyl digallate	3.33	± 0.44 [a]	27.80	± 4.88 [c]	10.47	± 0.50 [b]	ND	
Digalloylquinic acid	0.22	± 0.00 [a]	76.48	± 0.55 [d]	1.18	± 0.10 [b]	2.22	± 0.02 [c]
Gallotannins								
Trigalloylglucose	0.02	± 0.02 [a]	9.07	± 0.04 [d]	0.85	± 0.03 [b]	3.26	± 0.05 [c]
Tetragalloylglucose	1.64	± 0.02 [a]	14.50	± 0.89 [b]	3.16	± 0.13 [a]	88.37	± 3.22 [c]
Penta-galloylglucose	4.13	± 0.07 [a]	26.61	± 0.91 [b]	3.93	± 0.04 [a]	177.31	± 2.14 [c]
Hexagalloylglucose	4.00	± 0.08 [a]	74.46	± 0.03 [b]	9.78	± 0.25 [a]	63.72	± 7.92 [b]
Hepta-galloylglucose	3.59	± 0.00 [a]	58.26	± 1.75 [c]	7.75	± 0.15 [b]	ND	

Values with different superscripts [a,b,c,d] in the same row denoted significant differences according to the Student-Newman-Keulstest at $p < 0.05$. ND: not detected.

The occurrence of gallotannins, hydrolyzable tannins, was also detected in mango pulp and its by-products. Indeed, mango is recognized as one of the fruit with the major content of gallotannins [17]. Gallotannins contain gallic acid substituents esterified with glucose whose galloylation reaction yields tri-, tetra-, penta-, hexa-, and hepta-galloylglucoses (Table 1). Gallotannins concentrations varied according to the fraction of mango considered. Pulp and seed husk exhibited the poorest content of gallotannis (13.37 ± 0.02 and 25.47 ± 0.52 mg/100 $g_{dry\ matter}$, respectively) meanwhile peel and seed presented considerably higher total amounts (182.90 ± 3.62 and 332.66 ± 13.32 mg/100 $g_{dry\ matter}$, respectively). Mango peel was characterized by the significant concentrations of tri- and hepta-galloylglucose compared to the rest of the samples. On the other hand, mango seed resulted to be especially rich in tetra- and penta-galloylglucose. The high concentration of penta-galloylglucose in mango seed (177.31 ± 2.41mg/100 $g_{dry\ matter}$), which also was observed by other authors [13], should be noted due to its multiple functional properties such as antioxidant, anti-cancer, anti-viral, anti-microbial, anti-inflammatory, and anti-diabetic activities [18]. However, from a functional point of view, it should be taken into account that seed content in penta-galloylglucose could decrease drastically during the maturation process, as has been recently reported for mango pulp [19].

A large number of phenolic acid compounds derived from vanillic acid, hydroxybenzoic acid, coumaric acid, ferulic acid, and sinapic acid were also found in mango pulp and its by-products (Table 2). The most abundantly occurring phenolic acid derivative was vanillic acid glucoside whose concentrations in mango peel and seed were significant higher than in mango pulp and seed husk. Based on the results, mango peel exhibited the major amounts of phenolic acid derivative compounds. However, only the presence of vanillic acid glucoside, *p*-hydroxybenzoic acid glucoside, and ferulic acid hexoside on mango seed was confirmed. Therefore, phenolic acid derivative compounds seemed to not play an important role in the antiplatelet activity of mango seed extract.

Table 2. Mean concentration of gallic acid derivative compounds from different parts of *Keitt* mango variety at optimum maturation stage expressed as mg/100 $g_{dry\ matter}$.

COMPOUNDS	PULP		PEEL		SEED HUSK		SEED	
	MEAN	SD	MEAN	SD	MEAN	SD	MEAN	SD
Vanillic acid glucoside	3.90	± 0.19 [a]	16.68	± 0.42 [c]	7.94	± 0.36 [b]	15.91	± 0.77 [c]
p-Hydroxybenzoic acid glucoside	7.65	± 0.52 [c]	7.27	± 0.20 [c]	2.58	± 0.07 [a]	6.06	± 0.18 [b]
Dihydroxybenzoic acid glucoside	0.45	± 0.00 [a]	0.36	± 0.01 [a]	0.75	± 0.07 [b]	ND	
Hydroxybenzoyl galloyl glucoside	0.26	± 0.00 [b]	0.26	± 0.01 [b]	0.07	± 0.01 [a]	ND	
Coumaric acid glucoside	0.07	± 0.00 [a]	0.52	± 0.01 [c]	0.29	± 0.02 [b]	ND	
Coumaroyl galloyl glucoside	0.29	± 0.02 [a,b]	19.69	± 0.27 [c]	0.61	± 0.00 [b]	ND	
Ferulic acid hexoside	2.50	± 0.09 [b]	1.81	± 0.01 [a]	1.91	± 0.01 [a]	2.91	± 0.22 [c]
Sinapic acid hexoside	0.32	± 0.00 [a]	0.97	± 0.00 [b]	ND		ND	
Sinapic acid hexoside-pentoside	7.73	± 0.08 [b]	14.16	± 0.07 [c]	6.96	± 0.05 [a]	ND	
Dihydro sinapic acid hexoside-pentoside	3.57	± 0.07 [a]	4.90	± 0.08 [b]	3.53	± 0.09 [a]	ND	

Values with different superscripts [a,b,c] in the same row denoted significant differences according to the Student-Newman-Keuls test at $p < 0.05$. ND: not detected.

Other compounds of different nature and structure were also detected in mango samples (Table 3). Regarding ellagic acid and catechin, pulp and seed husk presented the lowest values, while seed was the mango fraction with the highest amount of ellagic acid ($7.46 ± 0.54$ mg/100 $g_{dry\ matter}$). The quantity of ellagic acid in mango seed extract had been reported from 3 to 156 mg equivalents of gallic acid per 100 g depending on the extraction method [20]. Several studies have proven the strong radical-scavenging activity of ellagic acid, even at very low concentrations [21,22]. Quercetin was also detected in glycosides, with the most common forms found being quercetin glucoside, quercetin galactoside, quercetin xyloside, and quercetin arabinopyranoside (Table 3). These flavonols were almost exclusively found in mango peel, which was in good agreement with findings reported by other authors [13,23]. Only small quantities of quercetin glucoside were detected in seed husk and seed.

Table 3. Mean concentration of ellagic acid, flavonols, xanthones, and benzophenones from different parts of *Keitt* mango variety at optimum maturation stage expressed as mg/100 $g_{dry\ matter}$.

COMPOUNDS	PULP		PEEL		SEED HUSK		SEED	
	MEAN	SD	MEAN	SD	MEAN	SD	MEAN	SD
Ellagic acid	0.02	± 0.00 [a]	1.35	± 0.03 [b]	0.56	± 0.04 [a]	7.46	± 0.54 [c]
Catechin	0.49	± 0.01 [a]	12.18	± 0.78 [c]	1.17	± 0.26 [a]	9.97	± 0.71 [b]
Quercetin glucoside	ND		21.28	± 0.63 [c]	0.21	± 0.05 [a]	1.39	± 0.14 [b]
Quercetin galactoside	ND		10.91	± 0.45 [b]	0.02	± 0.00 [a]	ND	
Quercetin xyloside	ND		2.83	± 0.25 [a]	ND		ND	
Quercetin arabinopyranoside	ND		2.94	± 0.20 [a]	ND		ND	
Rhamnetin hexoside	ND		2.37	± 0.05 [a]	ND		ND	
7-*O*-galloyltricetilflavan	ND		2.91	± 0.11 [a]	ND		13.99	± 0.29 [b]
Mangiferin	ND		ND		3.75	± 0.29 [a]	148.12	± 0.74 [b]
Maclurin C-glucoside	ND		ND		0.01	± 0.01 [a]	11.40	± 0.48 [b]
Maclurin galloyl glucoside	ND		4.60	± 0.04 [b]	0.84	± 0.12 [a]	9.63	± 0.04 [c]
Maclurin digalloyl glucoside	ND		1.63	± 0.04 [a]	ND		6.21	± 0.28 [b]
Iriflophenone glucoside	ND		ND		ND		10.89	± 0.28 [a]

Values with different superscripts [a,b,c] in the same row denoted significant differences according to the Studente-Newman-Keuls test at $p < 0.05$. ND: not detected.

On the other hand, the phenolic fraction of mango seed was characterized by the highest concentrations of 7-*O*-galloyltricetilflavan, mangiferin, and benzophenones such as maclurin glucoside, maclurin galloyl glucoside, maclurin digalloyl glucoside, and iriflophenone glucoside (Table 3). Among them, the most striking differences were those found on mangiferin content (1,3,6,7-tetrahydroxyxanthone-C2-β-D-glucoside). Mango seed showed the highest quantities of mangiferin ($148 ± 0.74$ mg/100 $g_{dry\ matter}$). Only small amounts of mangiferin were found in seed husk, while its occurrence was not detectable in pulp and peel. These results were in good agreement with those reported in bibliography, which also regarded mango seed as the part of mango with the highest mangiferin content [13,24]. Furthermore, the content of mangiferin in mango seed appeared to be influenced by the cultivar, as cv.*Keitt* presented major quantities of mangiferin in comparison with other varieties such as *Osteen* or *Sensación* [24]. The abundance of mangiferin in mango seed should be noted since mangiferin has been highlighted as a multi-target bioactive compound due to its health-endorsing properties. Indeed, several therapeutic and cosmetic applications have been recently attributed to mangiferin due to its antioxidative, antiaging, antidiabetic, anti-tumor, neuropropetective, cardiovascular, immunomodulatory and hepatoprotective effects, among others [25,26]. Consequently, the anti-platelet aggregation activity of mangiferin (1.0 mg mL^{-1}) on human platelet whose aggregation was induced by ADP (4 μM) was also estimated. Results showed that mangiferin has an important anti-platelet effect, exhibiting 31% inhibition (Figure 3). However, it has been demonstrated that for the same concentration, the antiplatelet aggregation activity increased more than two fold in mango seed. Therefore, although mangiferin showed a considerably bioactive effect, the antiplatelet activity of mango seed could also be explained by the action of other phenolic compounds and their possible synergistic interactions, or even by the presence of other compounds with reported bioactivity such as carbohydrates [27]. The percentage of carbohydrates reported in mango seed (64.24%) was considerably higher than that found in mango peel (31.24%) [28], which could have had an influence on the antiplatelet activity observed by mango seed. On the other hand, results compiled in bibliography also pointed out seed as the mango by-product with the highest antioxidant activity (28.92–32.61 g of Trolox equivalents per 100 g) compared to mango peel (5.39–6.01 g of Trolox equivalent per 100 g) evaluated by different in vitro methods [29]. Therefore, based our results and those found in other authors, mango seed seems to be an excellent source of phytochemicals with functional properties.

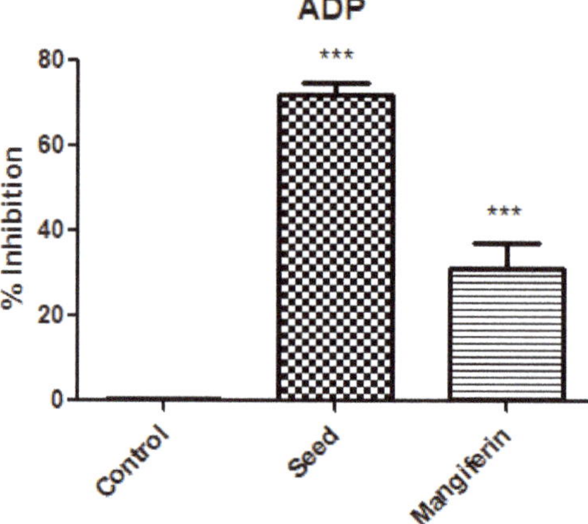

Figure 3. Antiplatelet aggregation activity results of mango seed extract and mangiferin at 1.0 mg/mL against ADP agonist (4 μM) expressed as mean value of inhibition percentages (*n* = 6). Data was analyzed using ANOVA of one factor. Post hoc analyses were conducted by Tukey's test. *** denotes significant differences compared to the negative control (absence of extract) at *p* = 0.001; ns denotes no statistical differences.

4. Conclusions

Mango seed was revealed as the fraction of the fruit with the most significant dose-dependent anti-platelet aggregation activity (inhibition percentage: 72%), compared to other mango by-products and the edible fraction. The chemical differences in phenolic composition found among different fractions of mango seemed to explain its bioactivity. Although mangiferin appeared to play a key role in this bioactivity, the antiplatelet effect of mango seed extract was not entirely explained by its action. Other compounds such as gallic acid, methylgallate and galloylquinic acid, tetra- and penta-galloylglucose, ellagic acid, 7-o-galloyltricetilflavan, iriflophenone glucoside, and maclurin C-glucoside and its derivatives were pointed out as contributors to antiplatelet aggregation activity. However, a comprehensive study of their individual antiplatelet properties should be addressed in further research.

Taken together, these results highlighted the use of mango seed as a promising natural co-product with antiplatelet properties, which can be used as a pharmaceutical drug or as a functional food ingredient with therapeutic applications against platelet aggregation.

Supplementary Materials: The following are available online at http://www.mdpi.com/2076-3921/8/11/517/s1, Table S1: Spectral and chromatographic data regarding to the bioactive compounds identified in mango and its by-products by HPLC-DAD-q-TOF-MS.

Author Contributions: Conceptualization, A.S.-C.; formal analysis, M.E.A.; investigation, M.E.A., L.R., E.F.; writing—original draft preparation, M.E.A.; writing—review and editing, I.P., D.A.-R., A.S.-C.; visualization, M.E.A.; supervision, D.A.-R., A.S.-C.; funding acquisition, I.P.; A.S.-C.

Funding: Ibero-American Program of Science and Technology for Development, Red iberoamericana de aprovechamiento integral de alimentos autóctonos subutilizados (ALSUB-CYTED, 118RT0543). Financial support of projects from program of aid for R + D + i of the Andalusian plan for research, development and innovation (PAIDI 2020) of Junta de Andalucía and Junta de Comunidades de Castilla-La Mancha (SBPLY/17/180501/000509).

Acknowledgments: M. Elena Alañón thanks to University of Castilla-La Mancha for the postdoctoral contract: Access to the Spanish System of Science, Technology and Innovation (SECTI). Authors are also grateful to the Company "Grupo Empresarial La Caña" for the mangoes traceability assurance and for its compromise with the research group and I + D + i.

Conflicts of Interest: The authors declare no conflict of interest.

References

1. Badimon, L.; Vilahur, G. Coronary Atherothrombotic Disease: Progress in Antiplatelet Therapy. *Revista Esp. de Cardiol.* (English Edition). **2008**, *5*, 501–513. [CrossRef]

2. Davies, M.J. The pathophysiology of acute coronary syndromes. *Heart* **2000**, *83*, 361–366. [CrossRef] [PubMed]

3. Geisler, T.; Anders, N.; Paterok, M.; Langer, H.; Stellos, K.; Lindemann, S.; Herdeg, C.; May, A.E.; Gawaz, M. Platelet response to clopidogrel is attenuated in diabetic patients undergoing coronary stent implantation. *Diabetes Care* **2007**, *30*, 372–374. [CrossRef] [PubMed]

4. Kong, Y.; Xu, C.; He, Z.L.; Zhou, Q.M.; Wang, J.B.; Li, Z.Y.; Ming, X. A novel peptide inhibitor of platelet aggregation from stiff silkworm, Bombyx batryticatus. *Peptides* **2014**, *53*, 70–78. [CrossRef] [PubMed]

5. Fuentes, E.; Palomo, I. Antiplatelet effects of natural bioactive compounds by multiple targets: Food and drug interactions. *J. Funct. Foods* **2014**, *6*, 73–81. [CrossRef]

6. Sagar, N.A.; Pareek, S.; Sharma, S.; Yahia, E.M.; Lobo, M.G. Fruit and vegetable waste: Bioactive compounds, their extraction and possible utilization. *Compr. Rev. Food Sci. F.* **2018**, *17*, 512–531. [CrossRef]

7. Muchiri, D.R.; Mahungu, S.M.; Gituanja, S.N. Studies on Mango (*Mangifera indica* L.) kernel fat of some Kenyan varieties in Meru. *J. Am. Oil Chem. Soc.* **2012**, *89*, 1567–1575. [CrossRef]

8. Ayala-Zavala, J.F.; Vega-Vega, V.; Rosas-Domínguez, C.; Palafox-Carlos, H.; Villa-Rodríguez, J.A.; Wasim Siddiqui, M.; Dávila-Aviña, J.E.; González-Aguilar, G.A. Agro-industrial potential of exotic by-products as a source of food additives. *Food Res. Int.* **2011**, *44*, 1866–1874. [CrossRef]

9. Jahurul, M.H.A.; Zaidul, I.S.M.; Ghafoor, K.; Al-Juhaimi, F.Y.; Nyam, K.-L.; Norulaini, N.A.N.; Sahena, F.; Mohd Omar, A.K. Mango (*Mangifera indica* L.) by-products and their valuable components: A review. *Food Chem.* **2015**, *183*, 173–180. [CrossRef]

10. Asif, A.; Farooq, U.; Akram, K.; Hayat, Z.; Shafi, A.; Sarfraz, F.; Sidhu, M.A.I.; Rehman, H.; Aftab, S. Therapeutic potentials of bioactive compounds from mango fruit waste. *Trends Food Sci. Tech.* **2016**, *53*, 102–112. [CrossRef]

11. Massibo, M.; He, Q. Major mango polyphenols and their potential significance to human health. *Compr. Rev. Food Sci. F.* **2008**, *7*, 309–319. [CrossRef]

12. Albuquerque Cavalacanti de Albuquerque, M.; Levit, R.; Beres, C.; Bedani, R.; De Moreno de LeBlanc, A.; Saad, S.M.I.; LeBlanc, J.G. Tropical fruit by-products water extracts as sources of soluble fibers and phenolic compounds with potential antioxidant, anti-inflammatory and functional properties. *J. Funct. Foods* **2019**, *52*, 724–733. [CrossRef]

13. Gómez-Caravaca, A.M.; López-Cobo, A.; Verardo, V.; Segura-Carretero, A.; Fernández-Gutiérrez, A. HPLC-DAD-q-TOF-MS as a powerful platform for the determination of phenolic and other polar compounds in the edible part of mango and its by-products (peel, seed and seed husk). *Electrophoresis* **2016**, *37*, 1072–1084. [CrossRef] [PubMed]

14. Fuentes, E.J.; Astudillo, L.A.; Gutierrez, M.I.; Contreras, S.O.; Bustamante, L.O.; Rubio, P.I.; Moore-Carrasco, R.; Alarcon, M.; Fuentes, A.; Gonzalez, J.A.; et al. Fractions of aqueous and methanolic extracts from tomato (*Solanum lycopersicum* L.) present platelet antiaggregant activity. *Blood Coagul. Fibrinolysis* **2012**, *23*, 109–117. [CrossRef]

15. Sai Srinivas, S.H.; Maneesh Kumar, M.; Prasada Rao, U.J.S. Studies on isolation and antioxidant properties of bioactive phytochemicals from mango peel harvested at different developmental stages. *Int. J. Agric. Environ. Bio-res.* **2017**, *2*, 136–148.

16. Siriamornpun, S.; Kaewseejan, N. Quality, bioactive compounds and antioxidant capacity of selected climateric fruits with relation to their maturity. *Sci. Hortic.* **2017**, *221*, 33–42. [CrossRef]

17. Smeriglio, A.; Barreca, D.; Bellocco, E.; Trombetta, D. Proanthocyanidins andhydrolyzable tannins: Occurrence, dietary intake and pharmacological effects. *Br. J. Pharmacol.* **2017**, *174*, 1244–1262. [CrossRef]

18. Torres-León, C.; Ventura-Sobrevilla, J.; Serna-Cock, L.; Ascacio-Valdés, J.A.; Contreras-Esquivel, J.; Aguilar, C.N. Penta-galloylglucose (PGG): A valuable phenolic compounds with functional properties. *J. Funct. Foods* **2017**, *37*, 176–189. [CrossRef]

19. Alañon, M.E.; Oliver-Simancas, R.; Gómez-Caravaca, A.M.; Arráez-Román, D.; Segura-Carrtero, A. Evolution of bioactive compounds of three mango cultivars (*Mangifera indica* L.) at different maturation stages analyzed by HPLC-DAD-q-TOF-MS. *Food Res. Int.* **2019**, *125*, 108526. [CrossRef]

20. Soong, Y.Y.; Barlow, P.J. Quantification of gallic acid and ellagic (*Mangifera indica* L.) kernel and their effects onantioxidant activity. *Food Chem.* **2006**, *97*, 524–530. [CrossRef]

21. Priyadarsini, K.I.; Khopde, S.M.; Kumar, S.S.; Mohan, H. Free radical studies fo ellagic acid, a natural phenolic antioxidants. *J. Agric. Food Chem.* **2002**, *50*, 2200–2206. [CrossRef] [PubMed]

22. Sroka, A.; Cisowski, W. Hydrogen peroxide scavenging, antioxidant and antiradical activity of some phenolic acids. *Food Chem. Toxicol.* **2003**, *41*, 753–758. [CrossRef]

23. Berardini, N.; Carle, R.; Schieber, A. Characterization of gallotannins and benzophenonederivates from mango (*Mangifera indica* L. cv. Tommy Atkins) peles, pulp and kernels by high-performance liquid chromatography/electrospray ionization mass spectrometry. *Rapid Commun. Mass Spectrom.* **2004**, *18*, 2208–2216. [CrossRef] [PubMed]

24. López-Cobo, A.; Verardo, V.; Díaz de Cerio, E.; Segura-Carretero, A.; Fernández-Gutiérrez, A.; Gómez-Caravaca, A. Use of HPLC- and GC-QTOF to determine hydrophilic and lipophilic phenols in mango fruit (*Mangifera indica* L.) and its by-products. *Food Res. Int.* **2017**, *100*, 423–434. [CrossRef] [PubMed]

25. Du, S.; Liu, H.; Lei, T.; Xie, X.; Wang, H.; He, X.; Tong, R.; Wang, Y. Mangiferin: An effective therapeutic agent against several disorders (Review). *Mol. Med. Rep.* **2018**, *18*, 4775–4786. [CrossRef] [PubMed]

26. Quadri, F.; Telang, M.; Mandhare, A. Therapeutic and socmetic applications of mangiferin: An updated patent review (patents published after 2013). *Expert Opin. Ther. Pat.* **2019**, *29*, 463–479. [CrossRef]

27. Sun, J.; Li, L.; You, X.; Li, C.; Zhang, E.; Li, Z.; Chen, G.; Peng, H. Phenolics and polysaccharides in major tropical fruits: Chemical compositions, analytical methods and bioactivities. *Anal. Methods* **2011**, *3*, 2212–2220. [CrossRef]

28. Tesfaye, T. Valorisation of mango fruit by-products: Physicochemical characterization and future prospect. *Chem. Process. Eng. Res.* **2017**, *50*, 22–34.

29. Nguyen, N.M.P.; Le, T.T.; Vissenaekens, H.; Gonzles, G.B.; Van Camp, J.; Smagghe, G.; Raes, K. In vitro antioxidant activity and phenolic profiles of tropical fruit by-products. *Int. J. Food Sci. Technol.* **2019**, *54*, 1169–1178. [CrossRef]

 *antioxidants*

Article

Polyphenolic Fraction from Olive Mill Wastewater: Scale-Up and in Vitro Studies for Ophthalmic Nutraceutical Applications

Maria Domenica Di Mauro [1,*], Giovanni Fava [1], Marcella Spampinato [1], Danilo Aleo [2], Barbara Melilli [2], Maria Grazia Saita [2], Giovanni Centonze [3], Riccardo Maggiore [3] and Nicola D'Antona [1,*]

[1] National Research Council of Italy, Institute of Biomolecular Chemistry (CNR–ICB), Via Paolo Gaifami 18, 95126 Catania, Italy; giovanni.fava@hotmail.com (G.F.); spampinatomarcella@gmail.com (M.S.)

[2] MEDIVIS, Corso Italia, 171, 95127 Catania, Italy; danilo.aleo@medivis.it (D.A.); barbara.melilli@medivis.it (B.M.); mariagrazia.saita@medivis.it (M.G.S.)

[3] Envisep srl–Z.Ind. VIII Strada, 29-95121 Catania, Italy; laboratorio@acimgroup.it (G.C.); riccardo.maggiore@libero.it (R.M.)

* Correspondence: mariadomenica.dimauro@virgilio.it (M.D.D.M.); nicola.dantona@icb.cnr.it (N.D.); Tel.: +39-095-733-8342 (M.D.D.M. & N.D.)

Received: 2 September 2019; Accepted: 4 October 2019; Published: 8 October 2019

Abstract: The valorization of food wastes is a challenging opportunity for a green, sustainable, and competitive development of industry. Approximately 30 million m^3 of olive mill wastewater (OMWW) are produced annually in the world as a by-product of the olive oil extraction process. In addition to being a serious environmental and economic issue because of their polluting load, OMWW can also represent a precious resource of high-added-value molecules such as polyphenols that show acclaimed antioxidant and anti-inflammatory activities and can find useful applications in the pharmaceutical industry. In particular, the possibility to develop novel nutraceutical ophthalmic formulations containing free radical scavengers would represent an important therapeutic opportunity for all inflammatory diseases of the ocular surface. In this work, different adsorbents were tested to selectively recover a fraction that is rich in polyphenols from OMWW. Afterward, cytotoxicity and antioxidant/anti-inflammatory activities of polyphenolic fraction were evaluated through in vitro tests. Our results showed that the fraction (0.01%) had no toxic effects and was able to protect cells against oxidant and inflammatory stimulus, reducing reactive oxygen species and TNF-α levels. Finally, a novel stable ophthalmic hydrogel containing a polyphenolic fraction (0.01%) was formulated and the technical and economic feasibility of the process at a pre-industrial level was investigated.

Keywords: olive mill wastewater; polyphenols; valorization; adsorbents; ophthalmic hydrogel; anti-inflammatory and antioxidant activity

1. Introduction

Olive mill wastewater (OMWW) is a complex mixture of vegetation waters, soft tissues of the olive fruit, and water used during the various stages of the olive oil extraction process, characterized by its dark color, strong odor, a mildly acidic pH, and a very high inorganic and organic load [1,2]; in particular, the organic content (biochemical oxygen demand BOD 35–132 g/L, chemical oxygen demand (COD) 30–320 g/L) [3] consists essentially of sugars, tannins, polyphenols, polyalcohols, proteins, organic acids, pectins, and lipids [4].

In general, the high polyphenolic content (0.5–24 g/L) [5] makes OMWW difficult to biodegrade and a serious environmental and economic issue. Several methods are reported in the literature concerning the treatment and disposal of OMWW such as anaerobic digestion, aerobic fermentation, and composting, but all of them involve the loss or destruction of many functional compounds [6–8]. On the other hand, polyphenolic compounds, well-known for their beneficial effects on human health, due to their antioxidant, cardioprotective, anticancer, anti-inflammatory, and antimicrobial properties [9,10] are nowadays widely recognized as valuable molecules in pharmaceutical and nutraceutical fields [11], and in such a context, OMWW represents a really challenging bioresource.

In particular, several studies have highlighted that hydroxytyrosol, the most abundant biophenol in OMWW acts as a free radical-scavenger and metal-chelator [12], protects against oxidative damage [13,14], inhibits the NADPH oxidase [15], the inducible form of nitric oxide synthase (iNOS), and the proinflammatory enzymes such as 5-lipoxygenase and cyclooxygenase [16], decreasing the production of nitric oxide, leukotrienes, and prostaglandins. Moreover, hydroxytyrosol is able to modulate the release of tumor necrosis factor-α (TNF-α) and other proinflammatory mediators [17,18].

Recent studies have reported that oxidative stress is involved in the pathogenesis of several eye diseases, such as ocular inflammation and dry eye disorder [19,20], therefore polyphenols could play an important role in the prevention and/or treatment of these pathologies [21].

Membrane separation, solvent extraction, resins treatment, centrifugation, chromatographic procedures, and enzymatic reactions [22–25] are different strategies reported in the literature to recover polyphenolic compounds from OMWW even if each of them presents a number of issues concerning both the industrial scalability and/or the economic and environmental sustainability.

In this paper we describe a novel route to selectively recover a polyphenolic fraction from OMWW, through a green and sustainable adsorption/desorption batch procedure, the scaling-up of the process at a pre-industrial level, and the in vitro studies to evaluate the cytotoxicity, antioxidant, and anti-inflammatory activities aimed to the formulation of a novel ophthalmic nutraceutical.

2. Materials and Methods

2.1. Materials

Fresh olive mill wastewater (OMWW) from Cerasuola cultivar were freshly collected in Menfi (Agrigento, Italy) in the middle-late of the olive oil processing season from a continuous three-phase olive oil processing mill operating at a malaxing temperature of 27 °C; samples were stored in airtight screw-capped tanks at −20 °C.

Folin–Ciocalteu reagent, trifluoroacetic acid (TFA), 3-[4,5-dimethyl(thiazol-2-yl)-3,5-diphenyltetrazolium bromide (MTT), 2′,7′-dichlorofluorescein diacetate (DCFH-DA), lipopolysaccharide (LPS) and all standards were purchased from Sigma-Aldrich (Milan, Italy). All solvents were purchased from Carlo Erba (Milan, Italy). The adsorbents Purosorb™PAD428, Purosorb™PAD550, and Purosorb™PAD900 were obtained from Purolite. Statens Seruminstitut Rabbit Cornea (SIRC) cells were purchased from the American Type Culture Collection (Rockville, MD, USA). Fetal Bovine Serum (FBS), Eagle's Minimum Essential Medium (EMEM), Phosphate Buffered Saline solution (PBS), and Penicillin-Streptomycin (5000 U/mL) were obtained from Gibco-BRL Life Technologies. *Escherichia coli* Formamidopyrimidine-DNA Glycosylase (Fpg) FLARE™ Module (4040-100-FM) was purchased from Trevigen. Rabbit Anti-TNF alpha antibody, Rabbit Anti-beta Actin antibody and Goat Anti-Rabbit IgG H&L (HRP) were purchased from Abcam. Super Signal West Pico Chemiluminescence detection system was purchased from Thermo Scientific (Rockford, IL, USA). The SkinEthic™ HCE (human corneal epithelium) tissues were purchased from Episkin (Lyon, France).

2.2. OMWW Pretreatment

Cerasuola-OMWW samples were centrifuged at 4000 rpm (2688 g) for 20 min to remove any solid residues of drupes and leaves and the supernatant was filtered through filter paper under vacuum condition as reported by Fava et al. (2017) [26]. Filtered OMWW were subjected to a flash-freezing process to avoid degradation of polyphenolic compounds and to ensure long-term stability and reproducibility of analyses. Samples stored at −20 °C into airtight screw-capped containers showed good stability for over 1 year; all analytical operations were performed, when possible, under argon or nitrogen as suggested by Obied et al. (2005) [27], and samples were treated preventing any alterations or contaminations by the environment.

2.3. Adsorption/Desorption Treatment

An aliquot of the chosen adsorbing material (Purosorb™PAD428, Purosorb™PAD900, and Purosorb™PAD550–10 g) was introduced in a column (3 × 50 cm), washed with a mixture of acetone/water (50/50) and then rinsed with water; bed column volumes amounted to 14 mL, 11 mL, and 15 mL respectively for Purosorb™PAD428, Purosorb™PAD900, and Purosorb™PAD550. The column was charged with filtered OMWW (10 mL) and eluted with pure water (50 ml) to collect the unabsorbed fraction. Subsequently, 50 mL of the chosen eluent was used to elute the column. Preliminarily different organic eluents or water/organic eluent mixtures were tested for polyphenols desorption, including methanol, ethanol, tetrahydrofuran, and ethyl acetate; in all cases, the best results were obtained with a water/ethanol (50/50) solution with a flow of 0.5 mL/min. The evaluation of maximum adsorption capacity for each resin was achieved by increasing the OMWW load volume. In order to be regenerated after use, the adsorbents were washed with ethanol (50 mL), dried, and kept at ambient temperature. Adsorbents were tested by consecutive adsorption/desorption cycles to define their recycling features.

2.4. Determination of Total Phenol Content

Total phenols were determined according to Di Mauro et al. (2017) [28]. Microplate spectrophotometer reader (Synergy HT multi-mode microplate reader, BioTek, Milano, Italy) was used to determinate the absorbance at λ 750 nm, and values compared against a gallic acid calibration curve ($y = 0.002x + 0.030$, $R^2 = 0.9997$). Results were expressed as g/L of gallic acid.

2.5. Determination of Total Carbohydrates

Total carbohydrates were determined by the Dubois method [29]. The absorbance values were evaluated at 490 nm and compared against a glucose calibration curve ($R^2 = 0.999$) from 1 to 100 mg/L (Cary UV Agilent Technology). Results were expressed as mg/L of glucose.

2.6. Determination of the Pollutant Load

Evaluation of the pollutant load of OMWW and eluted fractions was performed according to EPA (U.S. Environmental Protection Agency) methods by the following parameters: Chemical oxygen demand (COD, EPA test Method 410.3), biochemical oxygen demand (BOD5, EPA test Method 5210), total phosphorous (EPA test Method 365.3) and total nitrogen (EPA test Method 352.1).

2.7. Chromatographic Analysis of Polyphenols

HPLC-DAD (HITACHI) using a Kinetex C-18 (4.6 × 250 mm, 5 µM) column (Phenomenex) with a security guard cartridge (Phenomenex) was used for the analytic separation of polyphenols. The column temperature was maintained at 30 °C. Water (A) and acetonitrile (B) were used as eluents, both added with 0.1% trifluoroacetic acid (TFA). Samples were eluted according to the following gradient: 100% A as starting condition for 5 min; 58% A in 25 min; 100% B in 15 min, maintained for 5 min; flow rate 0.8 mL/min. Chromatograms were acquired at 280 nm. Identification of polyphenols was performed by comparison of UV spectra and retention times with the corresponding analytical standards: hydroxytyrosol, tyrosol, 1,2-dihydroxybenzene, oleuropein, 3-(4-hydroxyphenyl)propionic acid, syringic acid, 4-hydroxyphenylacetic acid (PHPA), 4-hydroxybenzoic acid, *p*-coumaric acid, *trans*-ferulic acid, gallic acid monohydrate, vanillic acid, caffeic acid, and verbascoside. For each commercial standard a 5-points calibration curve was used for the quantification.

2.8. H-NMR Analysis

Bruker Avance TM 400 spectrometer at 400.13 MHz was used to record ^{1}H-NMR spectra. All previously dried samples were solubilized in deuterium oxide (D_2O). Chemical shifts (δ) are given as parts per million relative to the residual solvent peak.

2.9. Adsorption Equilibrium Tests

Six different aliquots of the adsorbent material were weighted and placed in six beakers, to which 25 ml of OMWW were added. The quantities of adsorbent material were chosen in such a manner that the six beakers contained an adsorbent mass to OMWW volume ratio equal to 1:12, 1:10, 1:8, 1:6, 1:4, and 1:2, respectively. Thus the weighted masses of adsorbent were, in all cases, 2.08 g, 2.5 g, 3.12 g, 4.17 g, 6.25 g, and 12.5 g. The mixtures were stirred for 2 h and then let rest overnight, at 20 °C and in darkness, in order to reach the adsorption equilibrium. Samples were then filtered on 0.45 µM filters and analyzed for their content in Total Polyphenols and COD.

2.10. Adsorption Kinetic Tests

A weighted aliquot of adsorbent was placed in a beaker together with a proper volume of OMWW and subjected to continuous stirring. The adsorbent mass to the OMWW volume ratio, chosen on the base of previously obtained equilibrium adsorption data, was 10 g of adsorbent for 50 mL of OMWW. Small aliquots of the mixture (1 mL each) were sampled at predetermined time intervals, and analyzed for their total polyphenols content.

2.11. Cycling Efficiency Tests

A weighted aliquot of adsorbent was placed in a beaker together with a proper volume of OMWW and subjected to continuous stirring. The adsorbent mass to OMWW volume ratio, as in the kinetic tests, was 1:5. After a time of 30 min the mixture was filtered on a porous septum, the filtrate was collected, and the adsorbent was washed with water and placed again in the beaker, this time together with a hydroalcoholic solution (50% ethanol). After 15 min of stirring, this mixture was filtered, collected the filtrate phase, and then washed the adsorbent with water, to submit it to a new cycle. This sequence of operations was repeated 5 times. Total polyphenols concentration was determined in all filtrates, calculating the adsorption and desorption efficiency for each cycle.

2.12. Process Scale-Up

A prototype pilot plant was developed to scale up the laboratory process. Purosorb™PAD428 was used as an adsorbent phase since it gave the best results in the laboratory phase as described in the "Results and Discussion" section. Since residual oils and sand present in the input waters can cause problems of clogging, accumulation, and eroding or obstructing pipes and machinery, the first stage was the de-oiling and sand removal; a steel tank in which OMWW was released with a flow rate of 0.5 m^3/h and working simultaneously as a settler and oil-separator was used. De-oiled OMWW was then conveyed into a storage tank, and forced with a flow rate of 300 L/h, through a sand filter, in order to retain particles with a diameter >50 µM not intercepted in the previous sedimentation step.

Afterward, OMWW were subjected to an adsorption process on Purosorb™PAD428, with the aim to adsorb polyphenols; the system consisted of a steel vessel with a volume of 100 L, partially filled with 80 L (about 32 kg) Purosorb™PAD428. The filtered effluent flowed at a flow rate of 200 L/h, until 160 L of water was measured by a flowmeter, in accordance with the OMWW/ Purosorb™PAD428 ratio optimized in laboratory tests (5.0 ml OMWW/g stationary phase).

The obtained de-phenolized OMWW were addressed to a storage tank, which was used for different cosmeceutical applications [30]. After the adsorption phase, pure water was used to wash in countercurrent Purosorb™PAD428. Adsorbed polyphenols were than desorbed with 50 L water/ethanol/isopropanol 50:42:8 *v/v* (ethanol/isopropanol 85/15 used to prepare the eluent phase represented the purest commercially available composition for semi-industrial use) with a flow rate of 5 L/min, and stored in a 2000 L tank. As the last step, Purosorb™PAD428 was washed with 50 L ethanol/isopropanol 85/15, and then with water to eliminate alcoholic residues before restarting the cycle. Analytic control and chemical characterization on outputs from the various steps of the process was achieved by sampling points at various parts of the plant for each cycle sequence.

2.13. In Vitro Study

2.13.1. Cell Cultures and Treatments

SIRC cells (passage: 18) were cultured in a 12- or 96-multiwell microplate and/or in 25 cm^2 flasks, according to the type of assay, with EMEM supplemented with 1% penicillin-streptomycin and 10% fetal bovine serum, and incubated under a humidified atmosphere at 37 °C with 5% CO_2. When cells reached 70% of confluence, the lyophilized fraction PAD428-FR2 was dissolved in culture medium at the appropriate final concentrations for each biological assay.

2.13.2. MTT Assay

The potential cytotoxic effect of fraction was evaluated by MTT assay [30,31] on SIRC cells (2.5 × 10^4 cells/well) untreated and treated with different concentrations of PAD428-FR2 (0.01%, 0.02%, 0.05%, 0.1%) for 24 h; subsequently, 200 μL of MTT (0.5 mg mL^{-1}) in culture medium were added to each well and incubated for 3 h at 37 °C keeping a humidified atmosphere with 5% CO_2. Finally, the supernatant was aspirated off and 100 μL of DMSO was added to each well to dissolve the formazan crystals.

A microplate spectrophotometer reader (Synergy HT multi-mode microplate reader, BioTek, Milano, Italy) at λ = 550 nm was used to measure the optical density (OD). Results were expressed as a percentage of cell viability with respect to untreated control viable cells, whose value was equal to 100%.

2.13.3. Lactic Dehydrogenase Release

Lactic dehydrogenase (LDH) release was evaluated on SIRC cells untreated and treated with different concentrations of PAD428-FR2 (0.01%, 0.02%, 0.05%, 0.1%) for 24 h, measuring spectrophotometrically in the culture medium and in the cellular lysates, at λ = 340 nm by analyzing NADH reduction [32]. The percentage of lactic dehydrogenase release was calculated as the percentage of the total amount, considered as the sum of the enzymatic activity present in the cellular lysate and that in the culture medium.

2.13.4. Alkaline Comet Assay

The alkaline comet assay was performed on SIRC cells untreated and treated with PAD428-FR2 (0.01%, 0.02%, 0.05%) for 24 h. The minigels were prepared as described by Tomasello et al. (2017) [33]. Then, as described by Di Mauro et al. (2017) [34], the alkaline version of the comet assay was performed. An analysis of fifty nucleoids for each sample was carried on by using an epifluorescence microscope (Leica, Wetzlar, Germany) equipped with a camera. DNA damage was evaluated by using CASP (1.2.2) image analysis software. Results were expressed as the percentage of fragmented DNA present in the comet tail (%TDNA).

2.13.5. Reactive Oxygen Species (ROS) Determination

ROS levels were evaluated using 2',7'-dichlorofluorescein diacetate [35]. A microplate spectrofluorometer reader (Synergy HT multi-mode microplate reader, BioTek, Milano, Italy) was used to measure the fluorescence (λexcitation = 488 nm and λemission = 525 nm). The total protein content was evaluated for each sample according to Bradford (1976) [36]. The results were expressed as fluorescence intensity per mg protein.

2.13.6. Protective Effect against Oxidative Stress

SIRC cells were pretreated with PAD428-FR2 0.01% for 24 h and then stimulated with H_2O_2 (200 µM) for 10 and 30 min. Untreated cells were used as negative control; cells treated with H_2O_2 (200 µM) were used as a positive control. After treatments, cell viability was evaluated by MTT assay as previously described.

The oxidative DNA damage was evaluated by the Fpg-modified comet assay [37]. After lysis, each sample was incubated with 100 µL of enzyme dilution buffer or Fpg enzyme solution in a humidity chamber at 37 °C for 45 min. Then the samples were electrophoresed in alkaline solution (300 mM NaOH, 1 mM Na_2EDTA, pH > 13) for 20 min at 0.7 V/cm. After staining with SYBR green, the nucleoids were analyzed as already described.

ROS levels were evaluated using 2′,7′-dichlorofluorescein diacetate [35] as previously reported.

2.13.7. Protective Effect against Inflammation

SIRC cells were pretreated with PAD428-FR2 0.01% for 2 h and then stimulated with lipopolysaccharide (LPS 1 µg/mL) for 24 h. Untreated cells were used as negative control; cells treated with LPS (1 µg/mL) were used as a positive control. After treatments, cell viability was evaluated by MTT assay as previously described.

TNF-α protein expression was determined by Western Blot analysis. After treatments, proteins were extracted from SIRC cells as described by Anfuso et al. (2017) [38]. Protein samples (30 µg/lane) were subjected to SDS-PAGE and, after transferring to nitrocellulose membranes, were incubated with antibody against TNF-α and β-actin overnight at 4 °C followed by incubation with horseradish peroxidase conjugated secondary antibody, goat anti-rabbit for TNF-α and β-actin. Super Signal West Pico Chemiluminescence detection system was used to visualize the protein expression after washing with TBS-T. β-actin was used as the loading control. Image J software (Version1.43, Broken Symmetry Software, Bethesda, MD, USA) was used to analyze bands.

2.14. Studies for Ophthalmic Nutraceutical Application

2.14.1. Ophthalmic Formulation and Stability Study

PAD428-FR2 0.01% was formulated in a hydrogel whose composition is shown in Table 1.

Table 1. Composition of the ophthalmic hydrogel.

Ingredients	Composition (%)
PAD428-FR2	0.01
Carbopol®980	0.20
Pemulen™RT1-NF	0.01
EDTA	0.02
Glycerol	1.15
$Na_2HPO_4 \cdot 12H_2O$	0.25
NaOH	0.07
Purified water	Up to 100

Initially, three different solutions (A, B, and C) were prepared as described above. Solution A was prepared in an appropriate container (1 L) by dissolving 11.5 g of glycerol in 700 g of purified water. Next, 2 g of Carbopol®980 and 0.1 g of Pemulen™RT1-NF were added to the previous solution left under stirring for 2 h to ensure the complete dissolution of polymers. The resulting solution was sterilized in autoclave at 121 °C for 20 min. Solution B was prepared by dissolving 2.5 g of Na_2HPO_4 12 H_2O and 0.7 g of NaOH in 200g of purified water. Solution C was obtained by dissolving 0.2 g of EDTA and 0.10 g of lyophilized PAD428-FR2 in 79.7 g of purified water. Both solutions B and C were filtered through a hydrophilic filter PES Durapore 0.22 µM in a sterile environment and transferred into glass containers preliminarily sterilized. Solution B was added dropwise to the polymeric solution (A) stirring for 30 min. Finally, solution C was added to the obtained gel stirring for 60 min. The formulation was divided into 2 mL sterile polypropylene vials with screw cap and "O" silicone ring (Axygen®). All operations were carried out at 25 °C and in particular the process of filtration and closure was performed under sterile condition using a vertical laminar flow cabinet with HEPA filters (Bio Activa Plus, Aquaria®).

The pH, osmolality, appearance, and hydroxytyrosol concentration (%) were evaluated on different aliquots of hydrogel stored in dark conditions into 2 mL sterile polypropylene vials at 25 ± 2 °C with $60\% \pm 5\%$ relative humidity (R.H.) for 3 months.

2.14.2. Ocular Irritation Test

The ocular irritation potential of the formulation was evaluated by using SkinEthicTM HCE model according to the protocol and instructions of the manufacturer. Positive (ethanol-treated) and negative (PBS-treated) controls were used. The tissues were evaluated for cell viability (CV) using the MTT assay [39]. If the percentage of CV was >60%, the substance can be predicted as non-irritant (UN GHS classification: no category); if the percentage of CV was ≤60%, the substance can be considered irritant (UN GSH classification: Category 1 or Category 2).

2.15. Statistical Analysis

All results were obtained by three independent experiments performed in triplicate; the means and standard deviations for each value were calculated. One way ANOVA was used to assess statistical differences among different treatments. Bonferroni test was performed to obtain post hoc comparison. $p < 0.05$ as minimum level of significance was applied. Graph Prism version 5 and/or Microsoft Excel was used to perform all the analyses.

3. Results and Discussion

3.1. Selective Recovery of Polyphenolic Fraction from OMWW

Cerasuola-OMWW characterization was performed as described in previous works [26,28] and results are summarized in Table 2.

Table 2. Chemical characterization of olive mill wastewater (OMWW) and fractions.

Characterization	OMWW	PAD428-FR1	PAD900-FR1	PAD428-FR2	PAD900-FR2
Total Nitrogen (mg/L)	350.0 ± 3.4	318.0 ± 3.1	320.0 ± 3.2	1.0 ± 0.2	1.0 ± 0.2
Total Phosphorous (mg/L)	186.0 ± 2.1	144.0 ± 2.3	149.0 ± 2.1	0.5 ± 0.01	0.5 ± 0.01
COD (g/L)	73.65 ± 1.43	44.80 ± 1.34	44.02 ± 1.55	20.42 ± 1.25	21.07 ± 1.29
BOD5 (g/L)	38.44 ± 2.12	26.42 ± 2.05	26.40 ± 2.19	15.60 ± 1.99	15.33 ± 1.92
Total Polyphenols (g/L)	5.20 ± 0.14	0.40 ± 0.04	0.38 ± 0.04	4.12 ± 0.11	3.94 ± 0.12
Hydroxytyrosol (g/L)	1.10 ± 0.08	-	-	0.90 ± 0.06	0.85 ± 0.06
Tyrosol (g/L)	0.14 ± 0.02	-	-	0.10 ± 0.02	0.08 ± 0.01
Total Sugar (g/L)	34.00 ± 2.18	24.22 ± 1.14	23.75 ± 2.09	4.70 ± 0.59	4.82 ± 0.32

Mean values ($n = 9$) $\pm$ SD were calculated.

A valuable use of olive mill wastewaters could be obtained through the valorization of the recovered polyphenolic fraction. Practically, this is a hard task since biophenols are not very stable and can easily get oxidized, hydrolyzed, polymerized, conjugated, and/or complexated in this aqueous environment containing all the reactants (such as metals, enzymes, oxygen, and polysaccharides), required for these kinds of transformations. In a preliminary attempt to identify new, eco-sustainable, and cheaper materials, that are able to selectively adsorb the phenolic fraction (or a fraction of it), three "untraditional" materials were tested: corncob (the maize central core), coffee husk (a waste of the coffee torrefaction process), and volcanic ashes. These materials were chosen both because of their abundant presence in Sicily, and since they constitute wastes (in the case of volcanic ashes, wastes generated by nature itself), and were characterized by relevant disposal costs and/or polluting issues. All tested processes with these materials, regardless of the used conditions, did not give any significant results showing no selective or unselective adsorption property toward the compounds contained in OMWW. In a further attempt, three different commercial polymeric adsorbents (Purosorb™PAD428, Purosorb™PAD900, and Purosorb™PAD550) were tested separately to establish the most suitable to retain completely or partially the phenolic fraction in a selective way.

Different eluents and values of flow or temperature were tested to find a good compromise that would allow the ability to minimize possible oxidation reactions and achieve equilibria between the various phases. The general best conditions were a flow of 0.5 mL/min at 22 °C, and elution with water, to collect the unabsorbed fraction (FR1), followed by a solution of water/ethanol (50/50) in the attempt to selectively desorb the polyphenolic fraction (FR2), and finally with pure ethanol to wash and recycle the adsorbent phase. Analysis of different eluted fractions by HPLC-DAD indicated that only Purosorb™PAD428 and Purosorb™PAD900 were able to completely retain the phenolic compounds. In fact, both aqueous fractions (PAD428-FR1 and PAD900-FR1) presented a slight yellow color and were almost completely deprived of any phenolic component as demonstrated both by chromatograms at 280 nm and by ^{1}H-NMR spectra (data are shown in Supplementary materials Figures S2–S8); the results reported in Table 2 show that the eluted aqueous fractions (PAD428-FR1 and PAD900-FR1) had a very low amount of polyphenols (0.40 and 0.38 g/L). Furthermore, the values of COD and BOD were reduced by 40% and 30% respectively, and a slight decrease in the inorganic load was observed too (Table S1 shown in Supplementary material). On the other hand, high amounts of carbohydrates (not retained by the adsorbent resins) suggest that this fraction could find valuable applications in cosmetic field as described in a recent work [30].

Conversely, the fractions eluted with the ethanolic mixture (PAD428-FR2 and PAD900-FR2) presented an intense red–brown color and gave again similar results in terms of polyphenols content (up to 4.12 g/L, recovery 80%) as shown by the PAD428-FR2 sample chromatogram reported in Figure 1, with hydroxytyrosol and tyrosol respectively 0.90–0.85 g/L and 0.10–0.08 g/L as reported in Table 2. However other biophenols were identified in PAD428-FR2 samples as shown in Table S2 (see Supplementary materials).

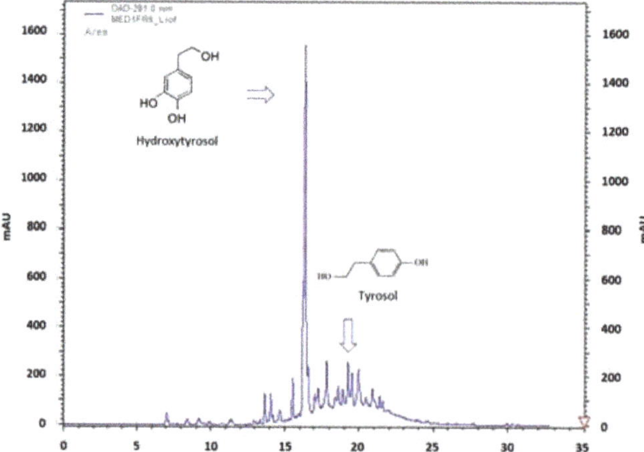

Figure 1. Chromatogram of PAD428-FR2 fraction at 280 nm.

The maximum volume of Cerasuola-OMWW that can be absorbed by 10 g of both adsorbent resins, before saturation, was 30 mL. Since Purosorb™PAD428 and Purosorb™PAD900 showed very similar adsorbent features, the subsequent adsorption equilibrium tests and scale-up process were carried out employing only the resin Purosorb™PAD428, and biological assays and formulation procedure of the ophthalmic nutraceutical were realized with the corresponding polyphenolic fraction (PAD428-FR2).

3.2. Adsorption Equilibrium and Kinetic Tests

To describe the equilibrium state in the adsorption system, the concept of dynamic equilibrium is commonly used. The liquid/gas molecules striking on the surface of a solid material can be adsorbed or rebounded; the rate of adsorption at the beginning is elevated since the adsorption sites are all available, but it decreases over time as the surface gets covered by adsorbate molecules. Conversely, the desorption rate increases because a greater number of molecules rebound until reaching the equilibrium between the adsorption rate and the desorption rate [40]. There are several equilibrium isotherm models but the most used and important in the field of adsorption for environmental cleanup are Freundlich and Langmuir isotherms [41,42].

After adsorption equilibrium was reached, batch test mixtures were filtered (on filter paper) and collected in test tubes. The resulting solutions for Purosorb™PAD428 resin test are shown in Figure 2a. The color diminution, that corresponds to the increase in the adsorbent dose (from right to left), was a clear symptom of the removal of the polyphenolic components (that are accounted for the typical reddish OMWW color). The solutions were analyzed, with respect to their content in COD and total polyphenols, and the results (see Figure 2b) confirmed the visual interpretation.

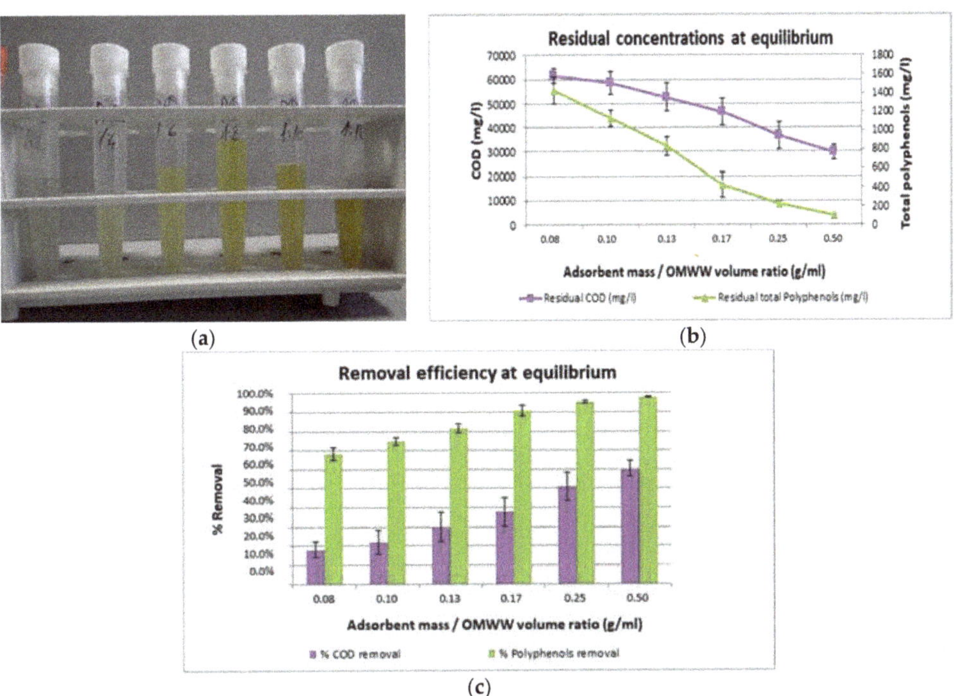

(a)

(b)

(c)

Figure 2. (a) Filtered solutions of OMWW, resulting from batch tests with Purosorb™PAD428 resin. (b) Residual concentrations of chemical oxygen demand (COD) and total polyphenols, at adsorption equilibrium, for the Purosorb™PAD428 resin. (c) Removal efficiency for COD and total polyphenols, at adsorption equilibrium, for the Purosorb™PAD428 resin. Mean values ($n = 5$) ± SD were calculated.

According to Figure 2c, a significant increase in total polyphenols removal efficiency was observed when the adsorbent dose to OMWW ratio increased from 1:12 to 1:2. The total polyphenols removal reached 90% at 1:6 ratio (0.17), while the removal rate increase was less pronounced at higher ratios. This result is easily understandable since an increase in the amount of adsorbent material will increase the total surface area and the available adsorption sites, lowering the driving force for intra-particle adsorption at each adsorption site, and resulting in a minor utilization of the adsorption capacity. This leads to the conclusion 1:5 ratio is considered to be a good compromise between removal efficiency and employed quantity of adsorbent.

The isotherm of total polyphenols adsorption onto Purosorb™PAD428 resin is depicted in Figure 3a, in which the R^2, associated with the Freundlich model fitting, is indicated. It is observed that the curve is concave downward, indicating a favorable isotherm leading to high adsorption capacity, as depicted in the values of Langmuir constants. In the fixed-bed adsorption column, a strongly favorable isotherm would also lead to a short mass transfer zone [43].

As reported in Figure 3b, a better representation of results is obtained by the Langmuir isotherm model. This was an expected result since, compared to Freundlich isotherm, this model is more flexible in modeling adsorption from highly concentrated water solutions. This result validates the assumption of monolayer homogenous adsorption of total polyphenols on polymeric resin.

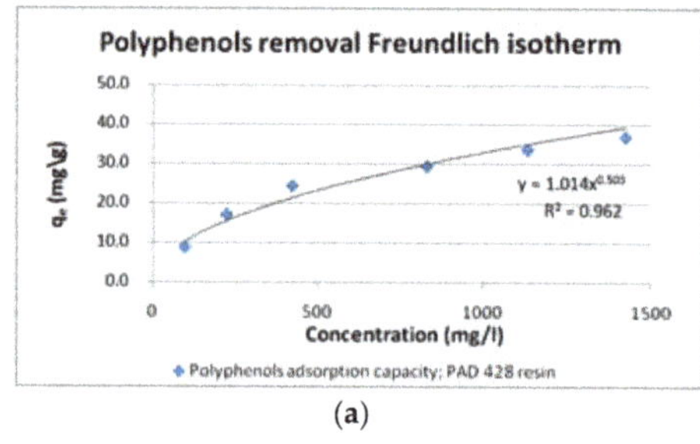

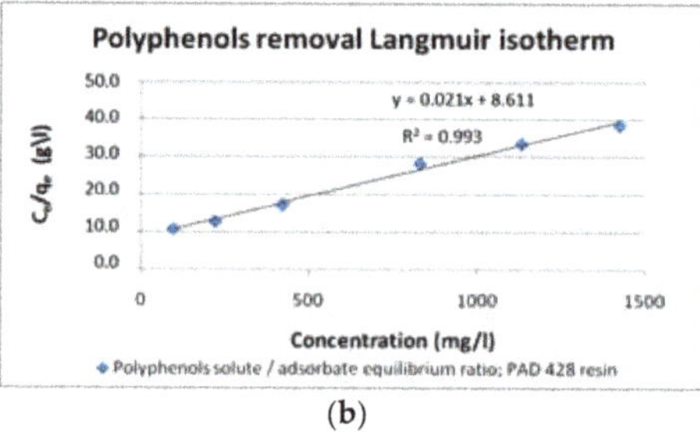

(a)	(b)

Figure 3. Freundlich (**a**) and Langmuir (**b**) isotherm curve for polyphenols adsorption, at equilibrium, 373 for the Purosorb™PAD428 resin. The y-axis reports: for the Freundlich curve, the equilibrium adsorption capacity "q_e" (q_e = x/m, in which x = mass of adsorbate and m = mass of adsorbent); for the Langmuir curve, the "C_e/q_e" ratio (in which C_e is the solute concentration at the equilibrium).

The correlation coefficient R^2 shown in Table 3 gives privilege for Langmuir isotherm over Freundlich one. The maximum Langmuir capacity Q^0 matches the equilibrium adsorption capacity curve (Figure 3b).

Table 3. Freundlich and Langmuir isotherms constants for total polyphenols adsorption on Purosorb™PAD428 resin.

Freundlich Isotherm			Langmuir Isotherm		
R^2	N	K_F	R^2	Q^0	K_L
		((mg/g)(L/mg) $^{1/n}$)			L/mg
0.962	2.0	1.0	0.993	47.7	0.02

Langmuir separation factor R_L resulted to be equal to 0.084, indicating that the adsorption isotherm of total phenol onto Purosorb™PAD428 resin is favorable ($0 < R_L < 1$). R_L values give evidence that the process is reversible, supporting the hypothesis that physical adsorption occurs in this process.

3.3. Cycling Efficiency Tests

One of the topics of adsorption systems is the number of cycles an adsorbent can undergo, which could be a limiting factor for the suitability and the overall economic convenience of the process.

Cycling tests were performed on Purosorb™PAD428 resin and Figure 4 shows results: it is confirmed, for the first cycle, that adsorption efficiency, after a 30 min contact time at 1:5 adsorbent mass to OMWW volume ratio, was about 91%, as found out in previous tests. After each cycle, this value slightly decreased down to about 79% after five cycles. Concerning the desorption efficiency values, these remained quite constant throughout the several cycles with values between 60–66%. Surprisingly, the incomplete polyphenols desorption from resin adsorption sites seems not to drastically influence its adsorption efficiency, maybe because it was charged at its maximum capacity during the tests. Anyway, this is an aspect that should be further investigated in future studies.

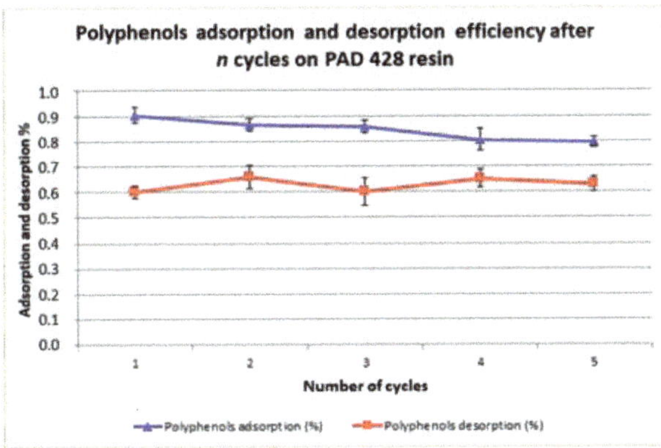

Figure 4. Polyphenols adsorption and desorption efficiency on Purosorb™PAD428 resin related to working cycle number. Mean values (n = 5) ± SD were calculated.

Figure 5 reports the overall polyphenols recovery. The recovery efficiency reduction, after five cycles, was of −7.5% (from 54% to 50%) for Purosorb™PAD428, even if a statistical analysis will be necessary for a better evaluation of data.

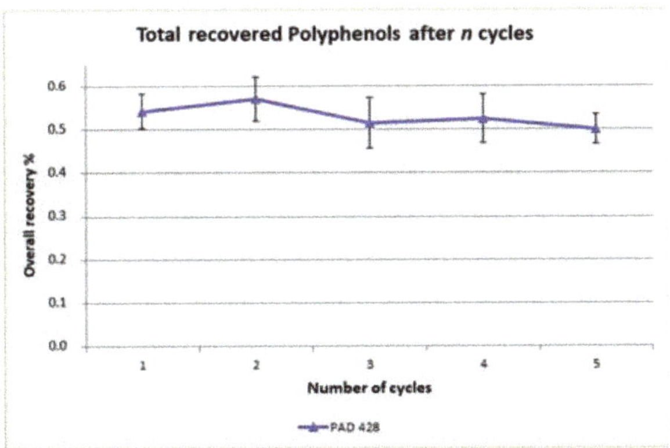

Figure 5. Total polyphenols recovered from OMWW using Purosorb™PAD428, related to the numbers of adsorption/desorption cycles. Mean values (n = 5) ± SD were calculated.

3.4. In Vitro Study and Ophthalmic Nutraceutical Application

Even if natural compounds and extracts are commonly considered free from harmful effects possible safety problems cannot be excluded [30]. Generally, irritation and cytotoxicity tests are performed to reduce the ocular risk of exposure to dangerous substances. Historically, animal tests such as the in vivo Draize rabbit assay were used to define the level of ocular toxicity by application of a test compound to a live rabbit's eye and then evaluation of the biological response. Recently, several in vitro alternative techniques have been developed as the result of ethical reconsideration of the animal used for toxicology studies [44].

For this purpose, we performed an in vitro study by using SIRC cells, a cell line having a mixed epithelial and fibroblastic nature [45], already used for toxicology studies in ophthalmic field [46,47].

Our preliminary interest was to evaluate in vitro the possible cytotoxic effect elicited by the fraction PAD428-FR2. Figure 6a shows the results of MTT assay, a colorimetric method measuring the reduction of MTT, a yellow-colored tetrazolium salt, to a purple formazan by the mitochondrial dehydrogenase enzyme of living cells [30,32]. The MTT assay data obtained on SIRC cells treated with different concentrations of PAD428-FR2 for 24 hours provided evidence to show that the lowest concentration (0.01%) did not influence the cell viability compared to untreated control cells. Conversely, the treatment with higher concentrations (0.02%, 0.05%, and 0.1%) caused a drastic decrease in cell viability in a dose-dependent manner with respect to both untreated control cells and 0.01% treated cells; in particular, these concentrations (0.02%, 0.05%, and 0.1%) were able to induce a dose-dependent necrotic effect as a result of cell membrane disruption (Figure 6b), in accordance with MTT assay data.

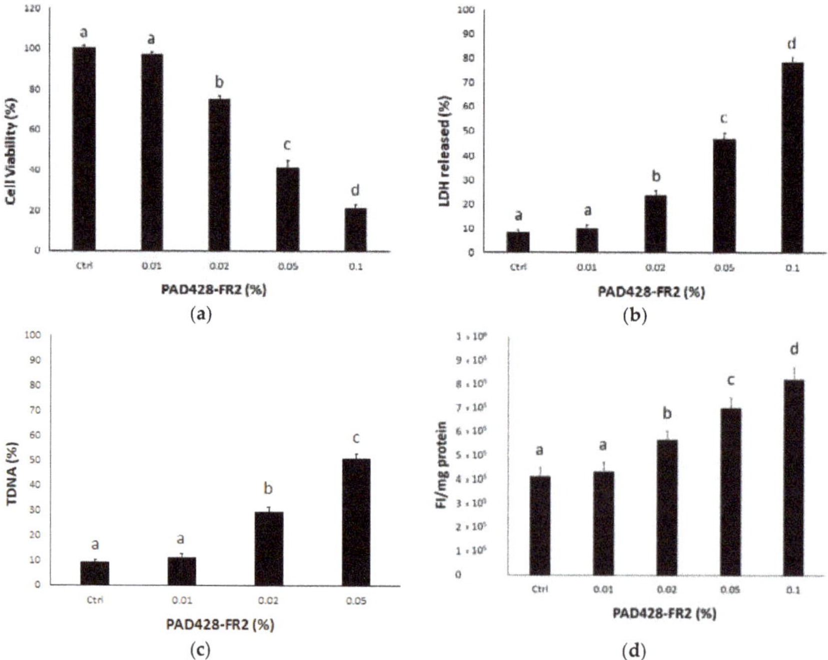

Figure 6. Effects of PAD428-FR2 on SIRC cells treated for 24 hours. (**a**) The results of 3-[4,5-dimethyl(thiazol-2-yl)-3,5-diphenyltetrazolium bromide (MTT) assay are expressed as the percentage of cell viability with respect to untreated control cells (Ctrl). (**b**) Lactic dehydrogenase (LDH) released. (**c**) DNA damage evaluated by alkaline comet assay; the results are expressed as the percentage of DNA present in the comet tail (%TDNA). (**d**) Reactive oxygen species (ROS) levels evaluated spectrofluorometrically by using 2′,7′-dichlorofluorescein diacetate (DCFH-DA); the results are expressed as fluorescence intensity (FI) per mg protein. Values are mean ± SD of three experiments in triplicate. Bars with different letters are significantly different ($p < 0.05$).

In addition, we evaluated in vitro the potential genotoxic effects of the fraction. Among the different approaches used to study DNA damage, comet assay is a simple and fast method to assess different types of DNA damage at the single-cell level [48]. In particular, we performed alkaline comet assay that permits to identify DNA double-strand breaks, single-strand breaks, alkali-labile sites, DNA-DNA/DNA-protein cross-linking, and incomplete excision repair sites [49]. The results, shown in Figure 6c, provided evidence that the treatment with 0.02% and 0.05% induced DNA damage in a dose-dependent manner.

In order to evaluate the possible mechanism of action, we measured ROS levels by using DCFH-DA, a non-fluorescent molecule that can spread through the cell membrane and, once inside the cell, is enzymatically hydrolyzed by an intracellular esterase to non-fluorescent DCFH; then ROS are able to oxidize DCFH to the fluorescent dichlorofluorescein (DCF), whose fluorescence intensity (FI) is proportional to the level of intracellular ROS [50].

In our opinion, the observed toxic effects could be related to the ROS overproduction (Figure 6d), that can determine the accumulation of oxidized intracellular macromolecules influencing the cell viability up until inducing cell death [51].

Based on these results we decided to continue the study by excluding the highest concentrations (ranging from 0.02% to 0.1%) because of their toxicity for the selected cell line.

It is well known that oxidative stress is involved in the pathogenesis of several eye diseases, such as ocular surface inflammation and dry eye disease [19,20]. So we evaluated, for the first time to our knowledge, the antioxidant and anti-inflammatory activities of polyphenolic fraction obtained from OMWW on SIRC cells.

To test the antioxidant activity, SIRC cells were pretreated with PAD428-FR2 0.01% for 24 h and then were stimulated with H_2O_2 (200 µM) for 10 and 30 minutes. The results of the MTT assay (Figure 7) showed that H_2O_2-treatment reduced cell viability compared to untreated control cell in a time-dependent manner. Moreover, it can be observed that the pretreatment with PAD428-FR2 0.01% determined an increase in cell viability with respect to H_2O_2-treated cells, suggesting a good protective activity of the fraction against the oxidant stimulus. In addition, we performed the alkaline version of the comet assay by using Fpg enzyme that recognizes and cuts the sites corresponding to oxidized guanine bases [37]. The results, shown in Figure 8, evidenced that the pretreatment with PAD428-FR2 0.01% was able to protect DNA from oxidative stress induced by H_2O_2 added for 30 min.

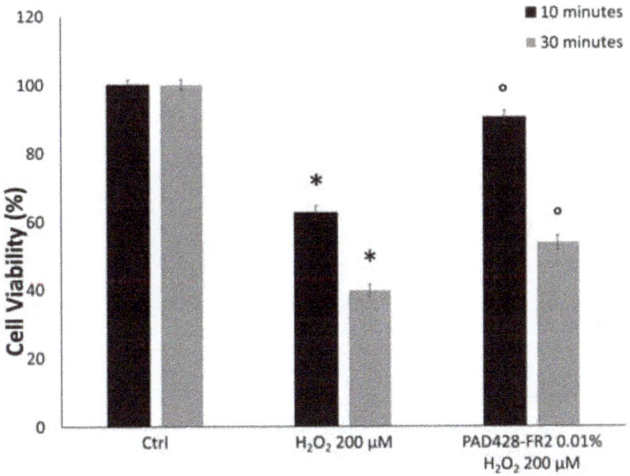

Figure 7. MTT assay performed on SIRC cells pretreated with PAD428-FR2 0.01% for 24 h and then stimulated with H_2O_2 (200 µM) for 10 and 30 min. Untreated cells were used as negative control; cells treated with H_2O_2 (200 µM) were used as a positive control. The results are expressed as the percentage of cell viability with respect to untreated control cells (Ctrl). Values are mean ± SD of three experiments in triplicate. * $p < 0.05$ vs. control group; ° $p < 0.05$ vs. H_2O_2-treated cells.

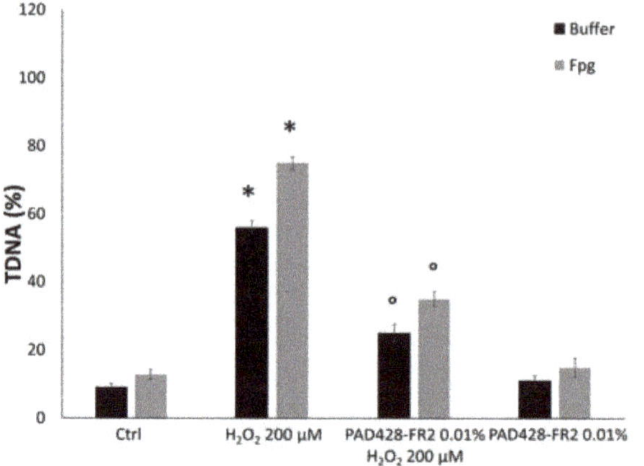

Figure 8. DNA oxidative damage evaluated by Fpg-modified comet assay performed on SIRC cells pretreated with PAD428-FR2 0.01% for 24 h and then stimulated with H_2O_2 (200 µM) for 30 min. Untreated cells were used as a negative control; cells treated with H_2O_2 (200 µM) were used as a positive control. The results are expressed as the percentage of DNA present in the comet tail (%TDNA). Values are mean ± SD of three experiments in triplicate. * $p < 0.05$ vs. control group; ° $p < 0.05$ vs. H_2O_2-treated cells.

As expected, a significant rise of ROS levels was observed in H_2O_2-treated cells with respect to untreated control cells in a time-dependent manner. This increase was counteracted by pre-incubating cells with PAD428-FR2 0.01% (Figure 9).

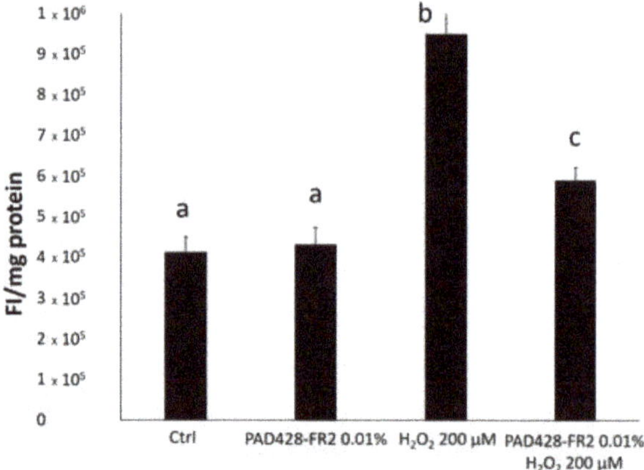

Figure 9. ROS levels evaluated on SIRC cells pretreated with PAD428-FR2 0.01% for 24 hours and then stimulated with H_2O_2 (200 µM) for 30 minutes. Untreated cells were used as a negative control; cells treated with H_2O_2 (200 µM) were used as a positive control. The results were expressed as fluorescence intensity per mg protein. Values are mean ± SD of three experiments in triplicate. Bars with different letters are significantly different ($p < 0.05$).

Our results are in accordance with Schlupp et al. (2019) who reported the antioxidant activity of a phenol-enriched OMWW extract was able to reduce the formation of free radicals in vitro [52]. Similar results were obtained by Schaffer et al. (2007) who demonstrated that hydroxytyrosol-rich olive mill wastewater extract was able to protect brain cells from oxidative damage [53].

Regarding the anti-inflammatory activity, SIRC cells were pretreated with PAD428-FR2 0.01% for 2 h and then were stimulated for 24 h with LPS (1 µg/ml), the major constituent of the outer membrane of Gram-negative bacteria able to elicit inflammation.

As expected, MTT data and immunoblots, shown in Figures 10 and 11, revealed that LPS-treatment significantly reduced the percentage of cell viability and increased the expression of TNF-α, a well-known pro-inflammatory mediator involved in ocular inflammation [38]. The results also revealed that the pretreatment with PAD428-FR2 0.01% determined a marked increase in cell viability and a decrease of TNF-α levels with respect to LPS-treated cells, demonstrating a good protective activity of fraction against the inflammatory stimulus.

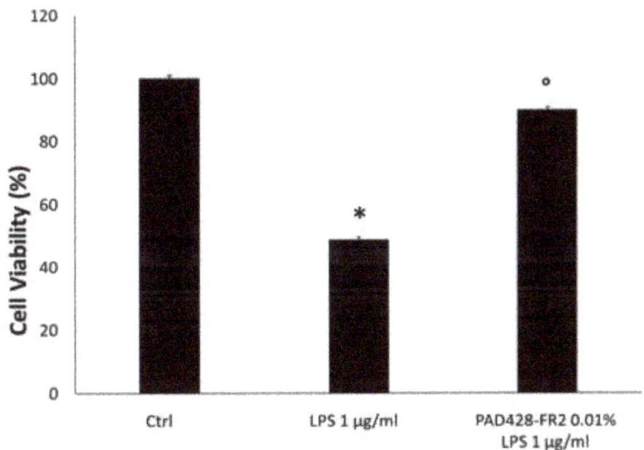

Figure 10. MTT assay performed on SIRC cells pretreated with PAD428-FR2 0.01% for 2 h and then stimulated with lipopolysaccharide (LPS) (1 µg/mL) for 24 h. Untreated cells were used as a negative control; cells treated with LPS (1 µg/mL) were used as a positive control. The results are expressed as the percentage of cell viability with respect to untreated control cells (Ctrl). Values are mean ± SD of three experiments in triplicate. * $p < 0.05$ vs. control group; ° $p < 0.05$ vs. LPS-treated cells.

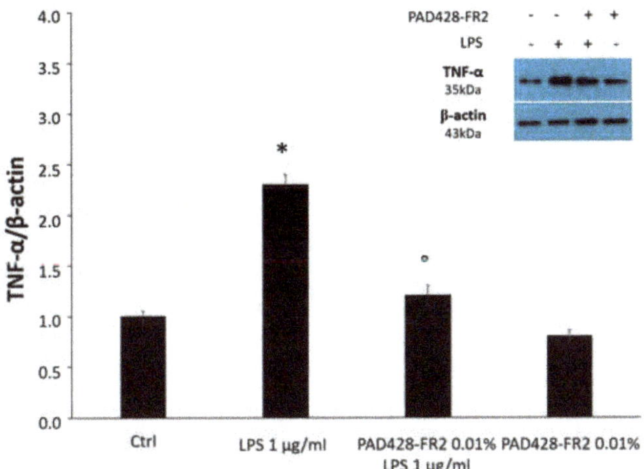

Figure 11. TNF-α levels evaluated by Western Blot analysis performed on SIRC cells pretreated with PAD428-FR2 0.01% for 2 h and then stimulated with LPS (1 µg/mL) for 24 h. Untreated cells were used as a negative control; cells treated with LPS (1 µg/mL) were used as a positive control. β-actin was used as the loading control. Values are mean ± SD of three experiments in triplicate. * $p < 0.05$ vs. control group; ° $p < 0.05$ vs. LPS-treated cells.

These data are in agreement with several in vitro and in vivo studies; for instance, Baci et al. (2019) reported that a polyphenol-rich extract from olive mill wastewater was able to induce a downregulation of pro-inflammatory pathways in prostate cancer cells [54]. Richard et al. (2011) evidenced that hydroxytyrosol was the major anti-inflammatory compound in aqueous olive extracts able to impair cytokine and chemokine production in murine macrophages stimulated with LPS [18]. In particular, Fuccelli et al. (2018) observed that hydroxytyrosol reduced the TNF-α secretion in LPS stimulated mouse model [17].

Based on the previous biological assays, 0.01% *w/w* of lyophilized PAD428-FR2 was used to formulate a novel nutraceutical ophthalmic preparation. Traditional ophthalmic liquid formulations are characterized by limited residence time in the eye due to lacrimal secretion and nasolacrimal drainage resulting in a low drug absorption (only 1–10%) and limited efficacy. The increase of viscosity of the formulation using biocompatible hydrophilic polymers with mucoadhesive properties represents one of the best strategies to prolong the residence time in the eye [55,56]. Indeed, several studies have reported that the water-base gels have several advantages over the traditional ophthalmic formulations, either in terms of enhanced therapeutic response or improved ocular bioavailability [57]. According to these considerations, in this study a combination of different hydrophilic polymers as Carbopol®980 and Pemulen™RT1-NF was employed to enhance the viscosity and obtain a hydrogel with a 3D polymeric network [58]. Hydrogel was prepared according to the protocol described in the "Materials and Methods" section, and the final composition is shown in Table 1.

The ophthalmic formulation was clear without any suspended particles or impurities and showed a pH value equal to 6.7 which is considered physiologically compatible. As reported in the Materials and Methods section, the values regarding pH, appearance, osmolality, and hydroxytyrosol (%) were evaluated to assess the chemical and physical stability in the storage period. The results of stability study, shown in Table 4, revealed no significant changes with respect to hydroxytyrosol concentration (%), pH, osmolality, and appearance in samples stored at 25 ± 2 °C with 60% ± 5% R.H. for 3 months.

Table 4. Stability study performed on ophthalmic hydrogel stored at 25 ± 2 °C with 60% ± 5% R.H. for 3 months.

	Months			
	0	1	2	3
Hydroxytyrosol (%)	100	99	98	98
pH	6.7 ± 0.1	6.7 ± 0.1	6.6 ± 0.1	6.7 ± 0.1
Osmolality	150.0 ± 1.0	153.0 ± 1.0	155.0 ± 1.0	155.0 ± 1.0
Appearance	transparent	transparent	transparent	transparent

Mean values ($n = 9$) ± SD were calculated.

Finally, we evaluated the ocular irritation of hydrogel using SkinEthicTM HCE model, composed of human corneal epithelial cells that forms a corneal epithelial tissue when cultured at the air–liquid interface in a chemically defined medium on a permeable synthetic membrane insert [39].

The percentage of cell viability evaluated by MTT assay was equal to 95.7 ± 5.3 for tissues treated with hydrogel. According to the viability classification prediction model, since the percentage of viability is more than 60%, the hydrogel can be classified as non-irritant.

3.5. Scaling up and Pilot Plant Development

A mandatory feature for a process aiming to be an applicative and possibly marketable development is economic sustainability. In order to evaluate this parameter and with the aim to test the scalability, reproducibility, and effectiveness of our protocols at a pre-industrial scale too, an automatized pilot plant for polyphenols extraction from OMWW was realized by using the optimized laboratory conditions.

In Figure 12 is the schematized pilot plant consisting of the following steps: de-oiling and sand removal through a settler and an oil-separator, sand filtration, adsorption process through Purosorb™PAD428 and discharging of the not retained dephenolized fraction (FR1), and finally desorption of the polyphenolic mixture (FR2) stored in a 2000 L accumulation tank by eluting with water\ethanol 50:50. A mixture of ethanol/isopropanol, 85:15, instead of pure ethanol, was used in the desorption phase because it represented the purest composition on the market for semi-industrial use.

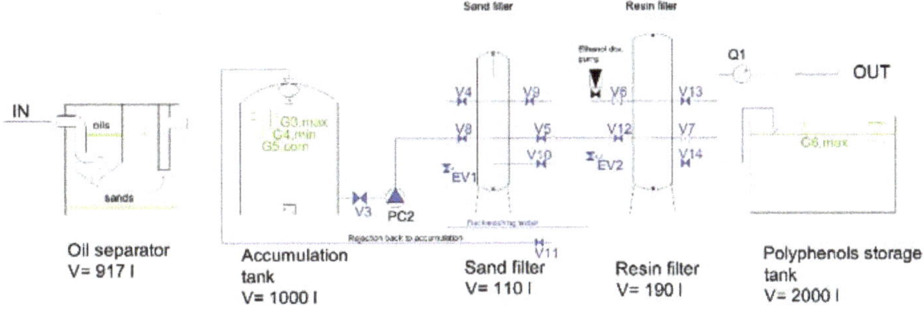

Figure 12. Scheme of the developed pilot plant prototype.

The empty bed contact time, with the operating flow, was about 48 min, a sufficient time to reach the maximum adsorption, according to laboratory results.

Preliminary runs in the pilot plant (see Figure S9 of prototype in Supplementary material) showed that the various fractions have features tending to satisfactorily match the laboratory results but need to be furtherly optimized. This is shown in Table 5, which compares olive mill wastewaters and fractions FR1/FR2 characteristics, obtained from the treatment in the pilot plant. It is to be noted that the difference in composition between lab and scale-up could partly be explained by the different purity of used eluents.

Table 5. Characterization of eluted fractions from preliminary pilot plant tests.

Characterization	OMWW	PAD428-FR1	PAD428-FR2
Total Nitrogen (mg/L)	350.0 ± 3.4	204 ± 1.6	1.0 ± 0.2
Total Phosphorous (mg/L)	186.0 ± 2.1	97 ± 1.4	0.5 ± 0.1
COD (g/L)	73.65 ± 1.43	50.60 ± 1.93	20.42 ± 1.64
BOD5 (g/L)	38.44 ± 2.12	19.82 ± 2.24	15.60 ± 1.45
Total Polyphenols (g/L)	5.20 ± 0.14	2.10 ± 0.09	1.12 ± 0.08
Hydroxytyrosol (g/L)	1.10 ± 0.08	not detected	0.45 ± 0.07
Tyrosol (g/L)	0.14 ± 0.02	not detected	0.06 ± 0.01
Total Sugar (g/L)	34.00 ± 2.18	23.50 ± 3.11	3.55 ± 3.38

Mean values ($n = 9$) ± SD were calculated.

The FR1 showed a reduced content in polyphenols (more than 50%), and a COD abatement of about 32%. This sugar-enriched fraction was used to develop novel cosmeceutical formulations as described in a previous work [30]; alternatively, it is possible to imagine for this fraction a final oxidative stage (aerobic microbial digestion) able to furnish clear depolluted water and, as a secondary product, some sludge that could find easy application as compost in agriculture due to its rich organic content. A further reduction of polyphenols content would help the potential subsequent biological sludge treatment, with the addition of urea and ammonium phosphate, to achieve the correct nutritional balance. After evaporation of the solvent, the residues of the FR2 alcoholic extract were analyzed by HPLC (chromatogram not reported): The main components, analogously to the lab-scale process, were tyrosol and hydroxytyrosol, with an average polyphenolic recovery of 22%, and a hydroxytyrosol purity of 40%. The lower polyphenolic recovery value (if compared to laboratory results) needs to be improved and investigations regarding the optimization of the scaling up parameters, especially flow rate and contact time, are actually in progress, but in general these firsts pilot plant operation experiments demonstrated that expected results are not too far from being reached.

In the pilot plant prototype, the total installed power (calculated as the sum of the nominal powers of all power-consuming sections) is about 1 kW, and considering a ten cycle operation, a preliminary economic balance was assessed, as illustrated in Table 6. Operational only costs were included, while any investment cost was excluded. Noteworthy, by recycling ethanol, that represents one of the major costs in the procedure, a substantial reduction of the total cost could be achieved, but one of the main objectives of subsequent studies should be a better regeneration of the resin and/or the extension of its useful life.

Table 6. Economical balance of the pilot prototype for ten-cycle operations.

Cost Item	Quantity (Units)	Unit Cost (€)	Total Item Cost (€)
Adsorbent (kg)	32	38.00	1216.00
Ethanol (L)	500	1.80	900.00
Manpower (hours)	2	25.00	50.00
Wastewater treatment (m^3)	2	0.80	1.60
Energy (kWh)	15	0.10	1.50
Mains water (m^3)	2	0.30	0.60
		Total 10-cycles cost	2169.70
Average polyphenolic extract production for 10 cycles (kg)			2
Estimated cost per kg of extract (€/kg)			1084.85

4. Conclusions

The described approach allows the ability to turn a highly polluting waste into valuable fractions, which can potentially be considered as raw starting materials for pharmaceutical/nutraceutical applications. A pilot plant prototype was realized and a preliminary economic balance calculated, showing the feasibility of the process at a pre-industrial level. In vitro biological assays were performed on the obtained polyphenolic fraction to study its cytotoxicity and anti-inflammatory/antioxidant activities, and the results permitted the ability to formulate a novel ophthalmic hydrogel with promising features that of course need yet to be more deeply characterized in view of a commercialization plan.

Supplementary Materials: The following are available online at http://www.mdpi.com/2076-3921/8/10/462/s1, Table S1: Metals analysis of Cerasuola-OMWW and fractions performed by ICP-MS, Table S2: Biophenols in Cerasuola-OMWW and PAD428-FR2 performed by HPLC-DAD analysis, Figure S1: Pictures of unabsorbed fraction eluted with water (PAD428-FR1, left) and fraction eluted with water/ethanol (50/50) solution (PAD428-FR2, right), Figure S2: Chromatogram of PAD428-FR1 at 280 nm, Figure S3: Chromatogram of PAD900-FR1 at 280 nm, Figure S4: Chromatogram of PAD550-FR1 at 280 nm, Figure S5: Chromatogram of PAD900-FR2 at 280 nm, Figure S6: Chromatogram of PAD550-FR2 at 280 nm, Figure S7: ^{1}H-NMR spectrum of PAD428-FR1 exhibited only signals in the typical regions of the alkylic (1–2 ppm) and heteroalkylic groups (3–4 ppm), confirming the absence of polyphenols and giving indications about the presence of carbohydrates, Figure S8: ^{1}H-NMR spectrum of PAD428-FR2 exhibited the characteristic signals of aromatic compounds (around 6.0–7.5 ppm) confirming the aromatic nature of the compounds in the mixture. Alkylic (1–2 ppm) and heteroalkylic groups (3–4 ppm) are still present and can be assigned to the sugar moiety of the glycosylated phenols, Figure S9: Picture of the pilot plant.

Author Contributions: Conceptualization, D.A., R.M., and N.D.; formal analysis, R.M.; funding acquisition, N.D.; investigation, M.D.D.M., G.F., M.S., and G.C.; project administration, N.D.; supervision, N.D.; validation, M.D.D.M., B.M., and M.G.S.; visualization, D.A.; writing—original draft, M.D.D.M. and N.D.; writing—review and editing, M.D.D.M. and N.D.

Funding: This research was funded by CNR (research project 1422 PO FESR Sicilia 2007–2013 linea d'intervento 4.1.1.1).

Acknowledgments: The authors wish to thank Sandro Dattilo and Emanuele Mirabella (National Research Council of Italy, Institute of Polymers, Composites, and Biomaterials, Via Paolo Gaifami 18, 95126 Catania, Italy) for the ICP-MS metal analysis (Supplementary material).

Conflicts of Interest: The authors declare no conflict of interest.

References

1. Deeb, A.A.; Fayyad, M.K.; Alawi, M.A. Separation of Polyphenols from Jordanian Olive Oil Mill Wastewater. *Chromatogr. Res. Int.* **2012**, *2012*, 812127. [CrossRef]
2. Kapellakis, I.E.; Tsagarakis, K.P.; Crowther, J.C. Olive oil history, production and by-product management. *Rev. Environ. Sci. Biol.* **2008**, *7*, 1–26. [CrossRef]
3. Dermeche, S.; Nadour, M.; Larroche, C.; Moulti-Mati, F.; Michaud, P. Olive mill wastes: Biochemical characterizations and valorization strategies. *Process Biochem.* **2013**, *48*, 1532–1552. [CrossRef]
4. Allouche, N.; Fki, I.; Sayadi, S. Toward a high yield recovery of antioxidants and purified hydroxytyrosol from olive mill wastewaters. *J. Agric. Food Chem.* **2004**, *52*, 267–273. [CrossRef] [PubMed]
5. D'Antuono, I.; Kontogianni, V.G.; Kotsiou, K.; Linsalata, V.; Logrieco, A.F.; Tasioula-Margari, M.; Cardinali, A. Polyphenolic characterization of Olive Mill WasteWaters, coming from Italian and Greek olive cultivars, after membrane technology. *Food Res. Int.* **2014**, *65*, 301–310. [CrossRef]
6. Fiorentino, A.; Gentili, A.; Isidori, M.; Lavorgna, M.; Parrella, A.; Temussi, F. Olive oil mill wastewater treatment using a chemical and biological approach. *J. Agric. Food Chem.* **2004**, *52*, 5151–5154. [CrossRef] [PubMed]
7. Günay, A.; Çetin, M. Determination of aerobic biodegradation kinetics of olive oil mill wastewater. *Int. Biodeter. Biodegr.* **2013**, *85*, 237–242. [CrossRef]
8. Papadimitriou, E.K.; Chatjipavlidis, I.; Balis, C. Application of composting to olive mill wastewater treatment. *Environ. Technol.* **1997**, *18*, 101–107. [CrossRef]
9. Fraga, C.G.; Croft, K.D.; Kennedy, D.O.; Tomás-Barberán, F.A. The effects of polyphenols and other bioactives on human health. *Food Funct.* **2019**, *10*, 514–528. [CrossRef]
10. Li, A.-N.; Li, S.; Zhang, Y.-J.; Xu, X.-R.; Chen, Y.-M.; Li, H.-B. Resources and biological activities of natural polyphenols. *Nutrients* **2014**, *6*, 6020–6047. [CrossRef]
11. Smiljković, M.; Dias, M.I.; Stojković, D.; Barros, L.; Bukvički, D.; Ferreira, I.C.F.R.; Soković, M. Characterization of phenolic compounds in tincture of edible Nepeta nuda: Development of antimicrobial mouthwash. *Food Funct.* **2018**, *9*, 5417–5425. [CrossRef] [PubMed]
12. Karković Marković, A.; Torić, J.; Barbarić, M.; Jakobušić Brala, C. Hydroxytyrosol, Tyrosol and Derivatives and Their Potential Effects on Human Health. *Molecules* **2019**, *24*, 2001. [CrossRef] [PubMed]
13. Grasso, S.; Siracusa, L.; Spatafora, C.; Renis, M.; Tringali, C. Hydroxytyrosol lipophilic analogues: Enzymatic synthesis, radical scavenging activity and DNA oxidative damage protection. *Bioorg. Chem.* **2007**, *35*, 137–152. [CrossRef] [PubMed]
14. Rodríguez-Ramiro, I.; Martín, M.A.; Ramos, S.; Bravo, L.; Goya, L. Olive oil hydroxytyrosol reduces toxicity evoked by acrylamide in human Caco-2 cells by preventing oxidative stress. *Toxicology* **2011**, *288*, 43–48. [CrossRef] [PubMed]
15. Maiuri, M.C.; De Stefano, D.; Di Meglio, P.; Irace, C.; Savarese, M.; Sacchi, R.; Cinelli, M.P.; Carnuccio, R. Hydroxytyrosol, a phenolic compound from virgin olive oil, prevents macrophage activation. *Naunyn Schmiedebergs Arch. Pharmacol.* **2005**, *371*, 457–465. [CrossRef]
16. Scoditti, E.; Calabriso, N.; Massaro, M.; Pellegrino, M.; Storelli, C.; Martines, G.; De Caterina, R.; Carluccio, M.A. Mediterranean diet polyphenols reduce inflammatory angiogenesis through MMP-9 and COX-2 inhibition in human vascular endothelial cells: A potentially protective mechanism in atherosclerotic vascular disease and cancer. *Arch. Biochem. Biophys.* **2012**, *527*, 81–89. [CrossRef] [PubMed]
17. Fuccelli, R.; Fabiani, R.; Rosignoli, P. Hydroxytyrosol Exerts Anti-Inflammatory and Anti-Oxidant Activities in a Mouse Model of Systemic Inflammation. *Molecules* **2018**, *23*, 3212. [CrossRef]
18. Richard, N.; Arnold, S.; Hoeller, U.; Kilpert, K.; Wertz, K.; Schwager, J. Hydroxytyrosol is the major anti-inflammatory compound in aqueous olive extracts and impairs cytokine and chemokine production in macrophages. *Planta Med.* **2011**, *77*, 1890–1897. [CrossRef]
19. Bucolo, C.; Fidilio, A.; Platania, C.B.M.; Geraci, F.; Lazzara, F.; Drago, F. Antioxidant and Osmoprotecting Activity of Taurine in Dry Eye Models. *J. Ocul. Pharmacol. Ther.* **2018**, *34*, 188–194. [CrossRef]
20. Dogru, M.; Kojima, T.; Simsek, C.; Tsubota, K. Potential Role of Oxidative Stress in Ocular Surface Inflammation and Dry Eye Disease. *Invest. Ophthalmol. Vis. Sci.* **2018**, *59*, DES163–DES168. [CrossRef]

21. Xu, Z.; Sun, T.; Li, W.; Sun, X. Inhibiting effects of dietary polyphenols on chronic eye diseases. *J. Funct. Foods* **2017**, *39*, 186–197. [CrossRef]

22. Khoufi, S.; Hamza, M.; Sayadi, S. Enzymatic hydrolysis of olive wastewater for hydroxytyrosol enrichment. *Bioresour. Technol.* **2011**, *102*, 9050–9058. [CrossRef]

23. Leouifoudi, I.; Zyad, A.; Amechrouq, A.; Oukerrou, M.A.; Mouse, H.A.; Mbarki, M. Identification and characterisation of phenolic compounds extracted from Moroccan olive mill wastewater. *Food Sci. Technol.* **2014**, *34*, 249–257. [CrossRef]

24. Leouifoudi, I.; Harnafi, H.; Zyad, A. Olive Mill Waste Extracts: Polyphenols Content, Antioxidant, and Antimicrobial Activities. *Adv. Pharmacol. Sci.* **2015**, *2015*, 714138. [CrossRef]

25. Zagklis, D.P.; Vavouraki, A.I.; Kornaros, M.E.; Paraskeva, C.A. Purification of olive mill wastewater phenols through membrane filtration and resin adsorption/desorption. *J. Hazard. Mater.* **2015**, *21*, 69–76. [CrossRef]

26. Fava, G.; Di Mauro, M.D.; Spampinato, M.; Biondi, D.; Gambera, G.; Centonze, G.; Maggiore, R.; D'Antona, N. Hydroxytyrosol Recovery from Olive Mill Wastewater: Process Optimization and Development of a Pilot Plant. *Clean Soil Air Water* **2017**, *45*, 1600042. [CrossRef]

27. Obied, H.K.; Allen, M.S.; Bedgood, D.R.; Prenzler, P.D.; Robards, K.; Stockmann, R. Bioactivity and Analysis of Biophenols Recovered from Olive Mill Waste. *J. Agric. Food Chem.* **2005**, *53*, 823–837. [CrossRef]

28. Di Mauro, M.D.; Giardina, R.C.; Fava, G.; Mirabella, E.F.; Acquaviva, R.; Renis, M.; D'Antona, N. Polyphenolic profile and antioxidant activity of olive mill wastewater from two Sicilian olive cultivars: Cerasuola and Nocellara etnea. *Eur. Food Res. Technol.* **2017**, *243*, 1895–1903. [CrossRef]

29. Dubois, M.; Gilles, K.A.; Hamilton, J.K.; Rebers, P.A.; Smith, F. Colorimetric method for determination of sugars and related substances. *Anal. Chem.* **1956**, *28*, 350–356. [CrossRef]

30. Di Mauro, M.D.; Tomasello, B.; Giardina, R.C.; Dattilo, S.; Mazzei, V.; Sinatra, F.; Caruso, M.; D'Antona, N.; Renis, M. Sugar and mineral enriched fraction from olive mill wastewater for promising cosmeceutical application: Characterization, in vitro and in vivo studies. *Food Funct.* **2017**, *8*, 4713–4722. [CrossRef]

31. Malfa, G.A.; Tomasello, B.; Acquaviva, R.; Genovese, C.; La Mantia, A.; Cammarata, F.P.; Ragusa, M.; Renis, M.; Di Giacomo, C. Betula etnensis Raf. (Betulaceae) Extract Induced HO-1 Expression and Ferroptosis Cell Death in Human Colon Cancer Cells. *Int. J. Mol. Sci.* **2019**, *20*, 2723. [CrossRef]

32. Malfa, G.A.; Tomasello, B.; Sinatra, F.; Villaggio, G.; Amenta, F.; Avola, R.; Renis, M. Reactive response evaluation of primary human astrocytes after methylmercury exposure. *J. Neurosci. Res.* **2014**, *92*, 95–103. [CrossRef]

33. Tomasello, B.; Malfa, G.A.; Strazzanti, A.; Gangi, S.; Di Giacomo, C.; Basile, F.; Renis, M. Effects of physical activity on systemic oxidative/DNA status in breast cancer survivors. *Oncol. Lett.* **2017**, *13*, 441–448. [CrossRef]

34. Di Mauro, M.D.; Ferrito, V.; Scifo, C.; Renis, M.; Tomasello, B. Effectiveness of Natural Compounds on DNA Damage in Coris julis (Linneaus 1758) from a Polluted Marine Area. *Water Air Soil Pollut.* **2017**, *228*, 228. [CrossRef]

35. Acquaviva, R.; Sorrenti, V.; Santangelo, R.; Cardile, V.; Tomasello, B.; Malfa, G.; Vanella, L.; Amodeo, A.; Mastrojeni, S.; Pugliese, M.; et al. Effects of extract of Celtis aetnensis (Tornab.) Strobl twigs in human colon cancer cell cultures. *Oncol. Rep.* **2016**, *36*, 2298–2304. [CrossRef]

36. Bradford, M.M. A rapid and sensitive method for the quantitation of microgram quantities of protein utilizing the principle of protein-dye binding. *Anal. Biochem.* **1976**, *72*, 248–254. [CrossRef]

37. Tomasello, B.; Grasso, S.; Malfa, G.; Stella, S.; Favetta, M.; Renis, M. Double-Face Activity of Resveratrol in Voluntary Runners: Assessment of DNA Damage by Comet Assay. *J. Med. Food* **2012**, *15*, 441–447. [CrossRef]

38. Anfuso, C.D.; Olivieri, M.; Fidilio, A.; Lupo, G.; Rusciano, D.; Pezzino, S.; Gagliano, C.; Drago, F.; Bucolo, C. Gabapentin Attenuates Ocular Inflammation: In vitro and In vivo Studies. *Front. Pharmacol.* **2017**, *8*, 173. [CrossRef]

39. Rodrigues, F.; Pereira, C.; Pimentel, F.B.; Alves, R.C.; Ferreira, M.; Sarmento, B.; Amaral, M.H.; Oliveira, M.B.P.P. Are coffee silverskin extracts safe for topical use? An in vitro and in vivo approach. *Ind Crop. Prod.* **2015**, *63*, 167–174. [CrossRef]

40. Bansal, R.C.; Goyal, M. *Activated Carbon Adsorption*; CRC Press: New York, NY, USA, 2005.

41. Freundlich, H.M. Over the Adsorption in Solution. *J. Phys. Chem. A* **1906**, *57*, 385–470.

42. Calisto, V.; Ferreira, C.I.; Oliveira, J.A.; Otero, M.; Esteves, V.I. Adsorptive removal of pharmaceuticals from water by commercial and waste-based carbons. *J. Environ. Manag.* **2015**, *152*, 83–90. [CrossRef]

43. McCabe, W.L.; Smith, J.C.; Harriot, P. *Unit Operations of Chemical Engineering*, 7th ed.; Chemical Engineering Series; McGraw-Hill: New York, NY, USA, 2005.

44. Wilson, S.L.; Ahearne, M.; Hopkinson, A. An overview of current techniques for ocular toxicity testing. *Toxicology* **2015**, *327*, 32–46. [CrossRef]

45. Olivieri, M.; Cristaldi, M.; Pezzino, S.; Rusciano, D.; Tomasello, B.; Anfuso, C.D.; Lupo, G. Phenotypic characterization of the SIRC (Statens Seruminstitut Rabbit Cornea) cell line reveals a mixed epithelial and fibroblastic nature. *Exp. Eye Res.* **2018**, *172*, 123–127. [CrossRef]

46. Cristaldi, M.; Olivieri, M.; Lupo, G.; Anfuso, C.D.; Pezzino, S.; Rusciano, D. N-hydroxymethylglycinate with EDTA is an efficient eye drop preservative with very low toxicity: An in vitro comparative study. *Cutan Ocul. Toxicol.* **2018**, *37*, 71–76. [CrossRef]

47. Wroblewska, K.; Kucinska, M.; Murias, M.; Lulek, J. Characterization of new eye drops with choline salicylate and assessment of their irritancy by in vitro short time exposure tests. *Saudi Pharm J.* **2015**, *23*, 407–412. [CrossRef]

48. Tomasello, B.; Malfa, G.; Galvano, F.; Renis, M. DNA damage in normal-weight obese syndrome measured by Comet assay. *Mediterr. J. Nutr. Metab.* **2011**, *4*, 99–104. [CrossRef]

49. Pu, X.; Wang, Z.; Klaunig, J.E. Alkaline Comet Assay for Assessing DNA Damage in Individual Cells. *Curr. Protoc. Toxicol.* **2015**, *65*, 1–11.

50. Rapisarda, V.; Loreto, C.; Ledda, C.; Musumeci, G.; Bracci, M.; Santarelli, L.; Renis, M.; Ferrante, M.; Cardile, V. Cytotoxicity, oxidative stress and genotoxicity induced by glass fibers on human alveolar epithelial cell line A549. *Toxicol. In Vitro* **2015**, *29*, 551–557. [CrossRef]

51. Muscoli, C.; Fresta, M.; Cardile, V.; Palumbo, M.; Renis, M.; Puglisi, G.; Paolino, D.; Nisticò, S.; Rotiroti, D.; Mollace, V. Ethanol-induced injury in rat primary cortical astrocytes involves oxidative stress: Effect of idebenone. *Neurosci. Lett.* **2002**, *329*, 21–24. [CrossRef]

52. Schlupp, P.; Schmidts, T.M.; Pössl, A.; Wildenhain, S.; Lo Franco, G.; Lo Franco, A.; Lo Franco, B. Effects of a Phenol-Enriched Purified Extract from Olive Mill Wastewater on Skin Cells. *Cosmetics* **2019**, *6*, 30. [CrossRef]

53. Schaffer, S.; Podstawa, M.; Visioli, F.; Bogani, P.; Müller, W.E.; Eckert, G.P. Hydroxytyrosol-rich olive mill wastewater extract protects brain cells in vitro and ex vivo. *J. Agric. Food Chem.* **2007**, *55*, 5043–5049. [CrossRef]

54. Baci, D.; Gallazzi, M.; Cascini, C.; Tramacere, M.; De Stefano, D.; Bruno, A.; Noonan, D.M.; Albini, A. Downregulation of Pro-Inflammatory and Pro-Angiogenic Pathways in Prostate Cancer Cells by a Polyphenol-Rich Extract from Olive Mill Wastewater. *Int. J. Mol. Sci.* **2019**, *20*, 307. [CrossRef]

55. Ludwig, A. The use of mucoadhesive polymers in ocular drug delivery. *Adv. Drug Deliv. Rev.* **2005**, *57*, 1595–1639. [CrossRef]

56. Almeida, H.; Amaral, M.H.; Lobão, P.; Sousa Lobo, J.M. In situ gelling systems: A strategy to improve the bioavailability of ophthalmic pharmaceutical formulations. *Drug Discov. Today* **2014**, *19*, 400–412. [CrossRef]

57. Zaki, R.; Hosny, K.M.; Khames, A.; Abd-elbary, A. Ketorolac tromethamine in-situ ocular hydrogel: Preparation, characterization and in vivo evaluation. *Int. J. Drug Del.* **2011**, *3*, 535–545.

58. Hoare, T.R.; Kohane, D.S. Hydrogels in drug delivery: Progress and challenges. *Polymer* **2008**, *49*, 1993–2007. [CrossRef]

 *antioxidants* 

Article

Recovery of Polyphenols from Brewer's Spent Grains

Rares I. Birsan [1,2], Peter Wilde [2], Keith W. Waldron [3] and Dilip K. Rai [1,*]

[1] Department of Food BioSciences, Teagasc Food Research Centre Ashtown, D15KN3K Dublin, Ireland
[2] Food Innovation and Health Programme, Quadram Institute Bioscience, Norwich Research Park, Colney NR4 7UQ, UK
[3] Anglia Science Writing Ltd., Wramplingham NR18 0RU, Norfolk, UK
[*] Correspondence: dilip.rai@teagasc.ie; Tel.: +353-(0)1-805-9500

Received: 30 June 2019; Accepted: 4 September 2019; Published: 7 September 2019

Abstract: The recovery of antioxidant polyphenols from light, dark and mix brewer's spent grain (BSG) using conventional maceration, microwave and ultrasound assisted extraction was investigated. Total polyphenols were measured in the crude (60% acetone), liquor extracts (saponified with 0.75% NaOH) and in their acidified ethyl acetate (EtOAc) partitioned fractions both by spectrophotometry involving Folin–Ciocalteu reagent and liquid-chromatography-tandem mass spectrometry (LC-MS/MS) methods. Irrespective of the extraction methods used, saponification of BSG yielded higher polyphenols than in the crude extracts. The EtOAc fractionations yielded the highest total phenolic content (TPC) ranging from 3.01 ± 0.19 to 4.71 ± 0.28 mg gallic acid equivalent per g of BSG dry weight. The corresponding total polyphenols quantified by LC-MS/MS ranged from 549.9 ± 41.5 to 2741.1 ± 5.2 µg/g of BSG dry weight. Microwave and ultrasound with the parameters and equipment used did not improve the total polyphenol yield when compared to the conventional maceration method. Furthermore, the spectrophotometric quantification of the liquors overestimated the TPC, while the LC-MS/MS quantification gave a closer representation of the total polyphenols in all the extracts. The total polyphenols were in the following order in the EtOAc fractions: BSG light > BSG Mix > BSG dark, and thus suggested BSG light as a sustainable, low cost source of natural antioxidants that may be tapped for applications in food and phytopharmaceutical industries.

Keywords: brewer's spent grain; polyphenols; microwave assisted extraction; ultrasound assisted extraction; liquid chromatography-mass spectrometry

---

1. Introduction

Brewers' spent grain (BSG) is generated in millions of tonnes every year as the major by-product of the brewing industry, with an annual global production estimated to be 39 million tonnes, of which EU generates ~8 million tonnes [1,2]. BSG is used as a low-value animal feed with a market value of ~35 Euro/tonne and thus making it an ideal substrate from which to recover high value compounds [3]. In addition to cellulose, hemicellulose, lignin, protein and lipids as the main components, BSG also contains low molecular weight phenolic compounds that have been associated with a wide array of health-benefiting properties [4,5].

A number of extraction methods, optimized and applied towards the recovery of polyphenols from BSG, have been comprehensively reviewed by several authors [3,6,7]. Depending on the types of BSG produced as a result of different cooking temperatures (70–250 °C), the polyphenol contents also differ between the lightly roasted malt producing light or pale BSG and the deeply roasted malts producing dark or black BSG. A common practice in breweries is to mix the light and dark malts in the ratio ~9:1 *w/w* in order to obtain the desired caramel colour and aroma of the beverage. Since BSG predominantly contains bound phenolics, chemical or enzymatic hydrolysis protocols are routinely used to release the phytochemicals bound to the cellular-wall components [8–10]. Solvent extraction or chemical hydrolysis combined with ultrasound (UAE) or microwave assisted extraction (MAE) or other physical cell-disruption techniques have been shown to increase the extraction yield of targeted compounds from BSG and similar biomass [7,11–13]. For example, in the recovery of BSG polyphenols, an optimised MAE method has been reported to result in a five-fold higher ferulic acid yield than the conventional solid–liquid extraction techniques [14]. In contrast, the same MAE parameters were also applied by Stefanello et al. [15] on BSG and corn silage, but the MAE yielded significantly lower total phenolic content than the conventional maceration method. In a separate study, mathematical models were used to optimize three extraction parameters (i.e., substrate to solvent ratio, extraction temperature and solvent composition) for MAE and UAE to recover maximum yield of unbound polyphenols from the unsaponified BSG. The subsequent experiments performed using the optimum parameters also resulted in higher polyphenolic contents by UAE (4.1 mg GAE/g BSG dw) and by MAE (3.9 mg GAE/g BSG dw) compared with the maceration method (3.6 mg GAE/g BSG dw) [15]. Both MAE, based on rapid heating of the solvent through microwave energy (that causes molecular motion via ionic conduction and dipole rotation), and UAE based on acoustic cavitation, increase the solvent penetration into the substrate leading to improved mass transfer rates. There is, however, a limited number of studies that focus on the UAE, MAE and conventional extraction methods to recover polyphenols from saponified BSG despite the presence of optimisation studies on individual methods in BSG [14–16] or similar substrates [17,18]. In addition, several of the aforementioned and other BSG polyphenol extraction studies were quantified spectrophotometrically using the Folin–Ciocalteu (FC) chemical method [15,16,19–22] either alone or with hyphenated chromatographic methods [15,16,20,22–26]. The spectrophotometric methods suffer generally by over estimating the phenolic contents since other non-polyphenolic molecules (e.g., reducing sugars) interact with the FC reagent used in the assay [27,28]. It is for this reason that in recent years researchers are discouraged from quantifying polyphenols using only the spectrophotometric methods [29,30].

In this study, we have investigated and compared the recovery of polyphenols from saponified light (L), dark (D) and BSG Mix using maceration, MAE and UAE techniques. The parameters for the various extraction methods have been adapted from the literature for maceration and UAE, whereas previously optimised parameters were applied for MAE. The objective of this study is to assess the polyphenol recovery from each type of BSG substrate using three different extraction methods. In addition, we have evaluated the enrichment of polyphenols through liquid-liquid partitioning of acidified 'liquors' (saponified fractions), which has been reported to lesser degree. Both the spectrophotometric and the LC-MS/MS methods have been employed and compared for the quantification of polyphenols in the various BSG fractions.

2. Materials and Methods

2.1. Materials and Chemicals

BSG L and D were provided by Diageo Ireland, Dublin. BSG Mix (light:dark, ~9:1 *w/w*) was obtained from the River Rye Brewing Company, Cellbridge, County Kildare, Ireland. The BSG samples were directly transported to the research centre within 30 min., oven-dried (Binder E28 oven, 72 h, 60 °C), milled (<1 mm) and vacuum packed until required.

The organic solvents (methanol, acetone, ethyl acetate (EtOAc), formic acid, acetonitrile), and sodium hydroxide (NaOH) were purchased from Merck (formerly Sigma Aldrich, Arklow, Co. Wicklow, Ireland). Polyphenol standards of gallic acid, *p*-coumaric acid, ferulic acid, sinapic acid, caffeic acid, protocatechuic acid, 4-hydroxybenzoic acid and +(-)catechin; and the chemicals FC reagent, hydrochloric acid and sodium carbonate were purchased from Merck (Arklow, Co. Wicklow, Ireland). Leucine-enkelphine was purchased from VWR International Ltd. (Blanchardstown, Dublin, Ireland).

2.2. Solid-Liquid Extraction

A schematic flow of the extraction procedure used is illustrated in Figure 1. Extraction of free (unbound) polyphenols referred to as crude extracts from BSG samples was carried out as in the previously optimised method [16], where 3 g milled BSG was mixed with 60 mL of 60% acetone at 60 °C for 30 min. with constant stirring. For the extraction of bound phenolics, 0.75% NaOH aqueous solution at 80 °C for 30 min. with constant stirring was used [14,18].

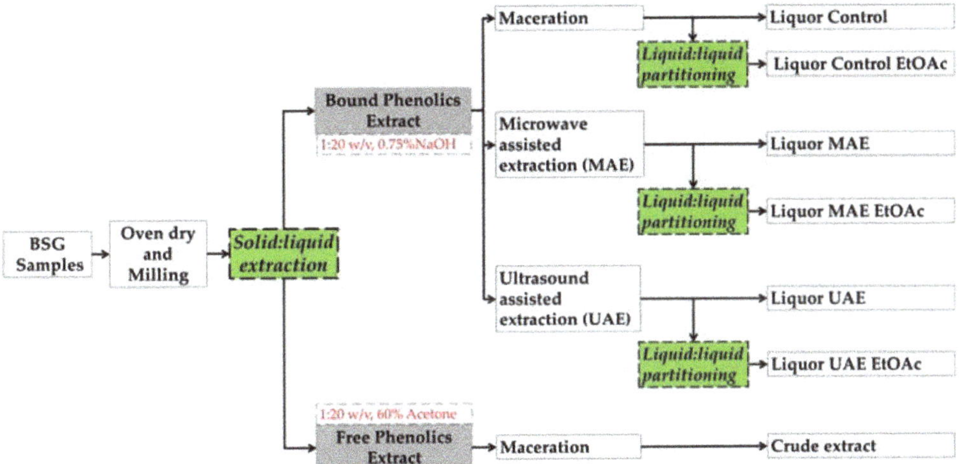

Figure 1. Flow chart showing the extraction procedure for brewers' spent grain (BSG) samples (light (L), dark (D), and Mix) for free phenolics and bound phenolics. Alkali-hydrolysed fractions (liquors) were partitioned with ethyl acetate (EtOAc).

2.3. Microwave Assisted Extraction

Microwave assisted extractions of BSG phenolics were performed according to the method previously optimized and reported by Moreira et al. [14]. The extraction was carried out in a microwave MARS™-6 (CEM, Matthews, NC, USA) equipped with a 40-position carousel. 2 g BSG samples were transferred to TFM extraction vessels with 40 mL alkali solution. Extraction was carried out for a duration of 15 min. at 100 °C. In all the vessels magnetic stirrers were added and used at maximum stirring speeds, while the pressure-leak and temperature were monitored for each vessel.

2.4. Ultrasound Assisted Extraction

Ultrasound assisted extraction was carried out on the Transonic TI-H-10 35 kHz sonication bath (ELMA Sch., Singen, Germany) at ~80 °C for 30 min. adapting the parameters previously optimised [17,18] in similar substrates. The substrate to solvent ratio (1:20 *w/v*) and the alkali concentration were maintained as used in the MAE and maceration methods, where 2.5 g BSG samples were mixed with 50 mL 0.75% NaOH solution in 100 mL amber bottles. The bottles were sealed to avoid any loss of solvents.

2.5. Preparation of Samples Following Maceration, MAE and UAE Treatments

After the extraction times were complete, all the extracts were left to cool at room temperature followed by centrifugation at 8400 rpm for 10 min. (MegaStar 600, VWR, Leuven, Belgium). The supernatants were pooled and syringe filtered through 0.45 μm PTFE filters for free phenolics, and PVDF filters for bound phenolic extracts. Aliquots (20 mL) of the liquor supernatants were acidified by adding hydrochloric acid solution (37%) until the pH reached 6.5 and subsequently subjected to liquid-liquid partitioning in EtOAc:water (1:1 *v/v*, 3 times) to obtain polyphenol-enriched fractions. The EtOAc fractions were evaporated to dryness under nitrogen and reconstituted in 20 mL 50% methanol. All the extractions were carried out in triplicate and stored at −25 °C until further use.

2.6. Total Phenolic Content Assay

Total phenolic content of BSG extracts was determined by colorimetric assay using FC reagent following [31]. Briefly, in 1.5 mL Eppendorf tube, 100 μL of extract was mixed with 100 μL each of methanol and FC reagent, and 700 μL of 20% sodium carbonate solution. The tubes were vortexed and incubated for 20 min. in darkness at room temperature. After the incubation, the tubes were centrifuged at 13,000 rpm for 3 min. to remove turbidity. Following this, 200 μL of the reaction mixture was transferred into 96-well micro plate and measured for absorbance at 735 nm using a spectrophotometer (FLUOstar Omega, BMG Labtech, Germany). Different concentrations of gallic acid as standards were used (10–300 μg/mL in 50% methanol) to prepare a calibration curve. The results are expressed in milligrams of gallic acid equivalent per gram dry weight (mg GAE/g BSG dw) BSG.

2.7. LC-MS/MS Identification and Quantification of BSG Phenolic Compounds

Quadrupole time-of-flight (Q-ToF) Premier mass spectrometer coupled to Alliance 2695 HPLC system (Waters Corporation, Milford, MA, USA) was used to profile various phytochemicals in the BSG L EtOAc fraction following the procedure previously described [32]. Accurate mass measurements of the molecular ions were achieved using an internal reference compound (Leucine–Enkephalin). The separation of the compounds was achieved on an Atlantis T3 C18 column (100 × 2.1 mm; 3 μm) using milliQ water (solvent A) and acetonitrile (solvent B) both containing 0.1% formic acid at a flow rate of 0.3 mL/min. at 40 °C. Electrospray ionisation (ESI) mass spectra were recorded on a negative ion mode for a mass range *m/z* 70–1000. Capillary and cone voltages were set at 3 kV and 30 V, respectively. Collision-induced dissociation (CID) of the analytes was performed using argon at 12–20 eV. Ultra-high performance liquid chromatography coupled to tandem quadrupole mass spectrometer (UPLC-TQD, Waters Corp., Milford, MA, USA) was used to quantify the BSG polyphenols by adapting the previous method used in raw barley [33]. Separation of the phenolics was carried out on an Acquity UPLC HSS T3 column (2.1 × 100 mm, 1.8 μm). The mobile phase consisted of milliQ water (solvent A) and acetonitrile (solvent B) both containing 0.1% formic acid. The UPLC separation was performed by an increasing organic solvent gradient from 2% to 98% B at a flow rate of 0.5 mL/min. for 10 min. The column temperature was set at 50 °C, while the samples were kept at 4 °C. The ESI source was set in negative mode and the quantification of each compound was performed using multiple reaction monitoring (MRM) method (Supplementary Table S1).

For the quantification of polyphenols, a stock solution (1000 ppm) for each standard was prepared and appropriate dilutions covering the range of 0.098 to 100 ppm were made to obtain standard curves. Targetlynx^TM integration software (Waters Corp., Milford, CT, USA) was used to quantify the compounds in the various extracts.

2.8. Statistical Analysis

Results are expressed as means of the triplicates ± standard deviation (SD). Differences between means were analysed using one-way analysis of variance with post-hoc Tukey test (SPSS Statistics 24). The statistical analysis on the different groupings was carried out using Minitab 18.0 (Minitab, Inc., State College, Pennsylvania, USA). The values were considered significantly different when $p < 0.05$.

3. Results and Discussion

3.1. Total Phenolic Content

The total phenolic content (TPC) from the crude extracts for the L, D and Mix BSG were 2.84 ± 0.11, 2.81 ± 0.14 and 3.85 ± 0.04 mg GAE/g BSG dw, respectively (Table 1). Past studies, by other authors, on the crude extracts of light and dark BSG have also shown TPC in a similar range [24,34]. These relatively low TPC levels in the crude extracts are because of the fact that the BSG contains a high amount of lignin ranging from 19.4–49.2 g/100 g that is connected to its cell wall polysaccharides by phenolic acids [10,16,35]. Therefore, it is essential to hydrolyse the rigid lignocellulose structural components to release the phenolic acids. Alkali hydrolysis is commonly used with BSG and other similar substrates. The TPC of the hydrolysed fraction (liquor) prior to acidification and partitioning is often reported, which is four- to five- times higher than the TPC values of the crude extracts [24,26]. For example, McCarthy et al. [25] recorded 16.0 mg GAE/g BSG dw and 18.3 mg GAE/g BSG dw for the light and dark BSG liquors, respectively. This trend is also evident from our study, where TPC values for the liquors ranged from 15.42 to 19.20 mg GAE/g BSG dw as opposed to the crude extracts (2.81 to 3.85 mg GAE/g BSG dw). Generally the dark BSG have shown higher levels of TPC values than the light BSG owing to the presence of high molecular weight melanoidins [20], which are accumulated as by-products of the Maillard reaction. The melanoidins mostly consist of sugar degradation products and amino acids [36] that can also react with FC reagent and thus give false elevated TPC.

Table 1. Total phenolic contents in mg GAE/g BSG dw in the NaOH saponified BSG extracts (liquors) and their subsequent ethyl acetate fractions following neutralisation (EtOAc); Ctrl represents maceration method, microwave assisted extraction (MAE), ultrasound assisted extraction (UAE) of light (L), dark (D) and Mix BSG. For each substrate, total phenolic content (TPC) values bearing different letters (a, b, c) are significantly different ($p < 0.05$) from each other. Shadow is to make the data distinguishable between the samples.

Samples	TPC mgGAE/g BSG dw		
	BSG L	**BSG D**	**BSG Mix**
Crude	2.84 ± 0.11 [c]	2.81 ± 0.26 [c]	3.85 ± 0.04 [c]
Liquor Ctrl	16.67 ± 0.87 [b]	17.27 ± 0.41 [ab]	19.20 ± 0.40 [a]
Liquor Ctrl EtOAc	4.67 ± 0.27 [c]	3.08 ± 0.15 [c]	4.71 ± 0.28 [c]
Liquor MAE	15.42 ± 1.16 [b]	15.55 ± 0.56 [b]	16.94 ± 1.84 [b]
Liquor MAE EtOAc	3.85 ± 0.19 [c]	3.01 ± 0.19 [c]	4.24 ± 0.22 [c]
Liquor UAE	15.76 ± 0.72 [b]	16.72 ± 0.96 [b]	16.99 ± 0.32 [b]
Liquor UAE EtOAc	4.17 ± 0.21 [c]	3.43 ± 0.46 [c]	4.62 ± 0.27 [c]

However, after the acidification of the liquors and subsequent partitioning with EtOAc, the TPC values of the EtOAc ranged between the crude and the liquor fractions (Table 1). Interestingly, the TPC of EtOAc fractions in the BSG D averaging 3.17 mg GAE/g dw is significantly lower than those of the L and Mix BSG averaging 4.23 and 4.52 mg GAE/g dw, respectively. Similar findings where the phenolics were lower in the hydrolysed dark BSG compared to light BSG have been reported by Moreira et al. [26]. Although the application of MAE and UAE techniques resulted, in general, lower TPC in the BSG EtOAc fractions than the conventional maceration method, but this decrease was not statistically significant except between the MAE and control BSG L. The possible reason for this decrease is due to the structural characteristic of the BSG as it predominantly contains a high lignin content [4,10]. It has been suggested before that the MAE is not able to promote sufficient molecular movement and rotation to overcome the lignin-barrier in contrast to constant stirring in the maceration method [14,15]. Furthermore the high temperature in MAE may induce the degradation of thermolabile polyphenols. A study on the effect of temperature on the extraction of polyphenols from *Gordonia axillaris*, an edible wild fruit, has shown a decrease in antioxidants' recovery with higher temperatures in MAE [37]. In general, high temperature has a positive effect on the extraction yield due to enhanced solubility and diffusivity of materials, however in UAE the high temperature has a negative effect on the extraction yield [38]. The high temperature increases the solvent vapour pressure and results in a decrease in surface tension that affect the cavitation bubble formation, which may explain the low TPC in the UAE treated samples.

3.2. LC-MS/MS Identification of BSG Polyphenols

As many as 14 different polyphenols were tentatively identified in the BSG L EtOAc extract using the accurate mass measurements, fragment ions and in conjunction with the literature (Figure 2, Table 2). Few of these polyphenols (protocatechuic acid and caffeic acid) were present in low amounts or co-eluted (syringic acid) with other phenolic acids as illustrated in the magnified inset in Figure 2 and the extracted ion chromatograms for these compounds in Supplementary Figure S1. Seven phenolic acids (ferulic acid, protocatechuic acid, 4-hydroxybenzoic acid, caffeic acid, syringic acid, p-coumaric acid and sinapic acid) and a flavonoid (catechin) were identified using commercially available standards and subsequently quantified using UPLC-TQD (Section 3.3). Several of the ferulic acid dimers and trimers listed in Table 2 have been identified previously in BSG using HPLC-DAD-MS/MS methods [4,14,39].

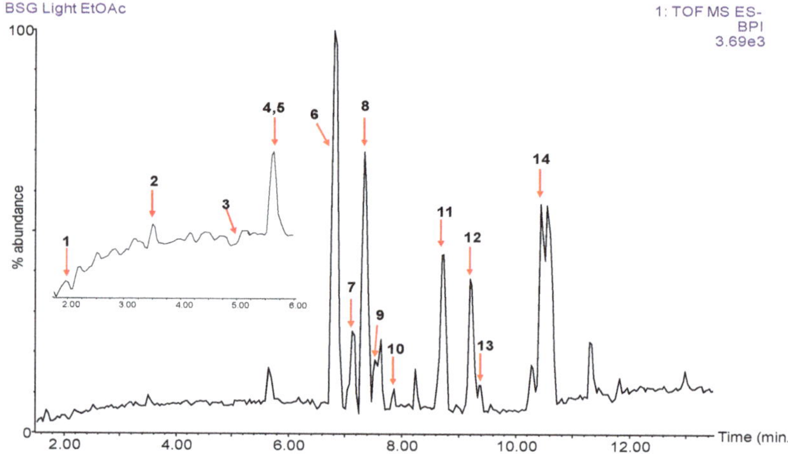

Figure 2. HPLC-Q-ToF (quadrupole time-of-flight) chromatogram of EtOAc fraction of BSG L showing the polyphenols (peaks 1–14) as assigned in Table 2. Shown in the inset is a close-up figure for the minor peaks 1–5. The elution time for peaks 1, 3 and 4 are demonstrated in their extracted ion chromatograms in Supplementary Figure S1.

Table 2. HPLC-Q-ToF identification of polyphenols in the ethyl acetate fraction of hydrolysed light BSG.

Peak No.	RT (min.)	Observed [M − H]⁻ (m/z)	Calculated [M − H]⁻ (m/z)	Chemical Formula	MS/MS Fragment Ions (m/z)	Tentative Identification
1	2.05	153.0169	153.0188	$C_7H_6O_4$	109.03	protocatechuic acid
2	3.50	137.0227	137.0239	$C_7H_6O_3$	93.04	hydroxybenzoic acid
3	4.93	179.0331	179.0344	$C_9H_8O_4$	135.04	caffeic acid
4	5.43	197.0452	197.0450	$C_9H_{10}O_5$	153.03	syringic acid
5	5.65	121.0282	121.0290	$C_7H_5O_2$	92.03	benzoic acid
6	6.80	163.0380	163.0395	$C_9H_8O_3$	119.05	coumaric acid
7	7.13	387.1073	387.1080	$C_{20}H_{20}O_8$	343.13, 193.05, 178.03, 149.07, 134.05	ferulic-ferulic acid dimer
8	7.34	223.0614	223.0606	$C_{27}H_{30}O_{16}$	179.02	sinapic acid
9	7.54	341.1019	341.1025	$C_{19}H_{18}O_6$	267.08, 193.05, 134.04	decarboxylated diferulic acid
10	7.87	385.0915	385.0923	$C_{20}H_{18}O_8$	282.09, 267.07 (100%), 239.08, 148.06	diferulic acid
11	8.73	385.0909	385.0923	$C_{20}H_{18}O_8$	325.09/326.09, 282.11/281.11 (100%), 267.08 (75%).	diferulic acid isomer
12	9.19	193.0516	193.0501	$C_{10}H_{10}O_4$	178.03, 134.04	ferulic acid
13	9.39	577.1342	577.1346	$C_{30}H_{26}O_{12}$	533.17, 355.09,	triferulic acid
14	10.44	341.1035	341.1025	$C_{19}H_{18}O_6$	326.09, 311.07, 282.09, 267.08 (100%), 239.08	decarboxylated diferulic acid isomer

In this study, an additional peak eluting at 7.13 min (peak 7) showed to contain a cluster of two molecules of ferulic acids corresponding to m/z 387.1073 [predicted molecular formula ($C_{20}H_{20}O_8$)]. On subjecting this molecular ion to MS/MS, the fragment ions m/z 343.1 (loss of CO_2), ferulic acid at m/z 193.1, m/z 149.1 (ferulic acid- CO_2) and m/z 134.0 (ferulic acid –(CO_2 CH_3)) further supported the detection of dimeric ferulic acid (Figure 3). Such non-covalent dimers generally form when the monomeric units are abundant in the sample, i.e., ferulic acid in this case.

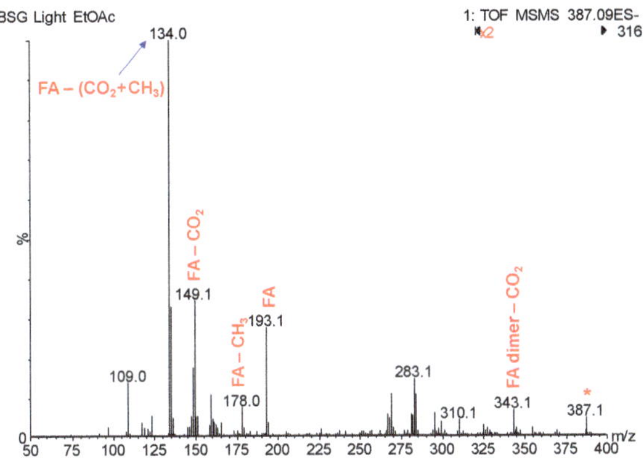

Figure 3. Electrospray ionisation (ESI)-MS/MS of m/z 387.1 showing the fingerprint fragment ions of the ferulic acid dimer. (FA = ferulic acid).

3.3. UPLC-MS/MS Quantification of BSG Polyphenols

Total polyphenols, the aggregate sum of individual polyphenols measured by UPLC-MS/MS, in each of the BSG EtOAc fractions, were found in decreasing levels of abundance in the following order: BSG L > BSG Mix > BSG D (Table 3). Statistically significant differences were found (in the same direction of abundance as TPC by FC) between the total polyphenols of BSG L, D and the Mix. The BSG L (2,741 µg/g dw) contained more than four times the total polyphenols found in BSG D (693 µg/g dw), which is in contrast to the TPC values where the dark BSG contained similar levels to light BSG as in this study (Table 1) or exceeded the light BSG [20,25]. The BSG Mix showed intermediate total polyphenol levels, i.e., between the BSG L and the BSG D as expected. Since BSG Mix constituted both the L and D (~9:1 *w/w*) BSG, we also measured the polyphenols in its crude and various 'liquor' fractions (prior to neutralisation and EtOAc partitioning) by UPLC-MS/MS. The crude extract of the BSG Mix contained low levels of polyphenols (~26 µg/g BSG dw), of which catechin constituted more than 50% of the total free polyphenols. This was 45- to 54- fold less than the total polyphenols present in the various EtOAc fractions (1170–1387 µg/g dw) of the same sample. McCarthy et al. [25] have also reported low levels of total polyphenols (30.6 µg/g in light and 27.2 µg/g in dark BSG dw) using HPLC coupled with diode array detector (DAD)-mass spectrometry analysis of the crude extracts. Stefanello et al. [15], on the other hand, have recorded 82.4 µg/g total polyphenols in the crude BSG extract, of which catechin constituted 83% of the total polyphenols. The TPC for these two studies ranged from 0.98–4.53 mg GAE/g BSG dw, which corroborate our findings. An even more interesting finding is that the total polyphenols in the liquors of BSG Mix were significantly lower than in the corresponding EtOAc fractions despite the fact that the TPC values for all 'liquor' fractions were very high (Tables 1 and 3). A similar observation was made by Stefanello et al. [15], where the TPC for the liquor was 17.4 mg GAE/g BSG dw, whilst the HPLC-DAD quantification of total polyphenols for the same liquor was 3195 µg/g dw. The HPLC-DAD value was closer to the TPC value of their crude BSG extract (3.43 mg GAE/g BSG dw). The high TPC values in the liquor fractions must have been attributed by other non-polyphenolic compounds such as reducing sugars, amino acids and peptides [4] that get fractionated in the water part during the EtOAc:water partitioning.

Table 3. UPLC-TQD quantification of BSG polyphenols *.

Samples	Ferulic Acid	*p*-Coumaric Acid	Catechin	4-Hydroxybenzoic Acid	Sinapic Acid	Syringic Acid	Protocatechuic Acid	Caffeic Acid	Total
BSG L Ctrl EtOAc	1809.5 ± 272.8 [a]	686.6 ± 59.0 [a]	2.11 ± 0.23 [b]	16.66 ± 4.45 [a]	14.63 ± 2.48 [a]	33.9 ± 10.44 [b]	3.46 ± 1.04 [ab]	0.147 ± 0.065 [d]	2741.1 ± 5.2 [a]
BSG L MAE EtOAc	1545.6 ± 157.3 [a]	499.1 ± 31.2 [bc]	1.43 ± 0.48 [b]	9.41 ± 1.15 [bcd]	11.02 ± 3.99 [ab]	18.9 ± 7.26 [bc]	1.38 ± 0.72 [cd]	0.370 ± 0.031 [b]	2087.2 ± 196.8 [a]
BSG L UAE EtOAc	1669.7 ± 21.8 [a]	579.2 ± 22.7 [b]	1.05 ± 0.07 [b]	10.76 ± 0.99 [bcd]	10.36 ± 1.52 [ab]	17.8 ± 3.68 [bc]	2.29 ± 0.83 [bc]	0.176 ± 0.013 [d]	2291.2 ± 42.7 [ab]
BSG D Ctrl EtOAc	404.7 ± 51.0 [cd]	185.3 ± 8.3 [f]	1.66 ± 1.01 [b]	13.12 ± 0.38 [ab]	7.63 ± 1.92 [bc]	76.4 ± 28.84 [a]	3.83 ± 0.63 [a]	0.407 ± 0.065 [b]	693.0 ± 85.7 [de]
BSG D MAE EtOAc	351.0 ± 33.9 [d]	155.3 ± 7.5 [f]	1.23 ± 0.33 [b]	11.36 ± 2.28 [bc]	4.68 ± 0.67 [c]	21.7 ± 4.84 [bc]	4.09 ± 0.55 [a]	0.547 ± 0.079 [a]	549.9 ± 41.5 [e]
BSG D UAE EtOAc	413.6 ± 135.8 [cd]	173.4 ± 56.6 [f]	2.18 ± 0.74 [b]	10.69 ± 1.39 [bcd]	8.28 ± 0.46 [bc]	17.3 ± 5.91 [bc]	4.85 ± 0.47 [a]	0.389 ± 0.052 [b]	629.9 ± 190.9 [de]
BSG Mix Ctrl EtOAc	894.6 ± 82.8 [b]	476.4 ± 35.1 [bcd]	nd	6.02 ± 0.93 [de]	9.59 ± 0.23 [abc]	nd	0.062 ± 0.012 [d]	0.226 ± 0.049 [cd]	1387.0 ± 119.0 [c]
BSG Mix MAE EtOAc	796.8 ± 68.1 [b]	355.4 ± 33.0 [e]	0.47 ± 0.82 [b]	6.88 ± 0.30 [cde]	10.23 ± 0.68 [ab]	nd	0.015 ± 0.026 [d]	nd	1169.8 ± 66.4 [c]
BSG Mix UAE EtOAc	848.5 ± 15.2 [b]	386.9 ± 6.7 [de]	nd	6.59 ± 0.55 [de]	11.33 ± 1.54 [ab]	nd	0.174 ± 0.085 [d]	0.328 ± 0.005 [bc]	1253.8 ± 11.3 [c]
BSG Mix Crude	2.8 ± 2.41 [e]	nd	14.05 ± 1.19 [a]	0.11 ± 0.12 [f]	8.28 ± 0.14 [bc]	nd	0.49 ± 0.17 [d]	nd	25.7 ± 1.97 [f]
BSG Mix Liquor Ctrl	714.1 ± 76.7 [bc]	423.3 ± 17.6 [cde]	1.09 ± 0.98 [b]	4.24 ± 0.50 [ef]	12.29 ± 1.09 [ab]	nd	nd	nd	1155.0 ± 93.2 [c]
BSG Mix Liquor MAE	647.4 ± 40.7 [bcd]	330.6 ± 49.5 [e]	1.86 ± 0.36 [b]	4.26 ± 0.33 [ef]	9.52 ± 0.29 [bc]	nd	nd	nd	993.6 ± 74.8 [cd]
BSG Mix Liquor UAE	739.1 ± 22.3 [b]	371.9 ± 30.9 [de]	nd	4.12 ± 0.37 [ef]	11.11 ± 0.^{39} [ab]	nd	nd	nd	1126.3 ± 53.2 [c]

* Values are expressed as µg/g BSG dw (mean ± SD); nd—not detected; For each substrate, the values reported, for each individual and total polyphenols in crude, liquors and their ethyl acetate (EtOAc) fractions bearing different letters (a, b, c, d, e, f) are significantly different ($p < 0.05$) from each other. Shadow is to make the data distinguishable between the samples.

In all the saponified BSG extracts, ferulic acid was the most predominant phenolic acid comprising in excess of 50% of the total polyphenols followed by *p*-coumaric acid. When the most abundant phenolic acid, i.e., ferulic acid is considered, there is no significant difference between the efficiency of the different extraction methodologies within the same type of BSG substrate. Several other studies have also established that the dominant polyphenols in BSG are ferulic acid and *p*-coumaric acid [10,26] and thus had become the target compounds of recovery in several studies [9–11,22,39–42]. Other abundant polyphenols in the BSG were sinapic acid and syringic acid, which have also been reported by other authors [25,42].

The UPLC-MS/MS determination of total polyphenols from MAE and UAE of the BSG EtOAc fractions showed a similar trend to their TPC values (Tables 1 and 3), where MAE and UAE yielded lower total polyphenols than the conventional maceration method. The lowest recovery of total polyphenols was by the MAE method. As explained earlier in Section 3.1, the MAE technique was not able to overcome the lignin-rich barrier, and that the extraction parameters used in the MAE and UAE may have induced thermal degradation of polyphenols.

The UPLC-MS/MS quantification of polyphenols in the various BSG EtOAc fractions was closer to the spectrophotometric FC-method (Table 3 vs. Table 1). Athanasios et al. [43] have used gas chromatography-mass spectrometry (GC-MS) and showed total polyphenols ranged between 2688 to 4884 µg/g dw in the four different batches of BSG, although the authors did not perform spectrophotometric analysis but these values are very close to TPC values of BSG in general.

4. Conclusions

The UPLC-MS/MS data have shown that the saponification followed by acidification and subsequent liquid-liquid partitioning (EtOAc) is the best procedure for polyphenol recovery and enrichment from BSG irrespective of extraction method. Without neutralisation and partitioning, the colourimetic chemical method falsely overestimates the total phenolic content and levels quantified by related assays in the liquors. Hyphenated chromatographic quantification methods such as LC-MS/MS is therefore necessary to accurately portray levels of total BSG polyphenols present. UAE and MAE treatments did not improve the BSG polyphenol yield indicating the thermal degradation of polyphenols with the extraction parameters used in these systems. The findings also suggest that ultrasonic bath operating at 35 kHz is less efficient in aqueous solution for the extraction of polyphenols from BSG. However, these techniques may improve the polyphenol yield and efficacy with further optimisation and when used with other systems, such as ultrasonic probes, and in combination with appropriate organic solvents.

Supplementary Materials: The following is available online at http://www.mdpi.com/2076-3921/8/9/380/s1, Table S1: Multiple reaction monitoring (MRM) transitions, cone voltages and collision energies used for the UPLC-TQD quantification of BSG polyphenols. Figure S1: Extraction ion chromatograms for peak 1 (*m/z* 153.017 [M − H]⁻), peak 3 (*m/z* 179.0133 [M − H]⁻) and peak 4 (*m/z* 197.045 [M − H]⁻).

Author Contributions: D.K.R. designed the experiments and interpreted the data; R.I.B. performed the experiments, analysed the data, and wrote the paper; K.W.W. and P.W. contributed to the design of experiments, interpretation of the data and writing the paper.

Funding: This research was funded by Teagasc Walsh Fellowship, grant number 2014027.

Acknowledgments: The authors would like thank Diageo Ireland, Dublin and the Rye River Brewing, Celbridge, Co. Kildare for kindly providing the brewer spent grains.

Conflicts of Interest: The authors declare no conflict of interest.

References

1. Statista. Beer Production Worldwide from 1998 to 2017 (in Billion Hectoliters). 2019. Available online: https://www.statista.com/statistics/270275/worldwide-beer-production/ (accessed on 20 February 2018).

2.   TBOE. Beer Statistics—2018 Edition. 2019. Available online: https://brewersofeurope.org/site/media-centre/key-facts-figures.php (accessed on 4 June 2019).

3.   Lynch, K.M.; Steffen, E.J.; Arendt, E.K. Brewers' spent grain: A review with an emphasis on food and health. *J. Inst. Brew.* **2016**, *122*, 553–568. [CrossRef]

4.   Jay, A.J.; Parker, M.L.; Faulks, R.; Husband, F.; Wilde, P.; Smith, A.C.; Faulds, C.B.; Waldron, K.W. A systematic micro-dissection of brewers' spent grain. *J. Cereal Sci.* **2008**, *47*, 357–364. [CrossRef]

5.   Shahidi, F.; Yeo, J. Bioactivities of phenolics by focusing on suppression of chronic diseases: A review. *Int. J. Mol. Sci.* **2018**, *19*, 1573. [CrossRef] [PubMed]

6.   Brglez Mojzer, E.; Knez Hrnčič, M.; Škerget, M.; Knez, Ž.; Bren, U. Polyphenols: Extraction methods, antioxidative action, bioavailability and anticarcinogenic effects. *Molecules* **2016**, *21*, 901. [CrossRef] [PubMed]

7.   Guido, L.F.; Moreira, M.M. Techniques for extraction of brewer's spent grain polyphenols: A review. *Food Bioprocess Technol.* **2017**, *10*, 1192–1209. [CrossRef]

8.   Krygier, K.; Sosulski, F.; Hogge, L. Free, esterified, and insoluble-bound phenolic acids. 1. Extraction and purification procedure. *J. Agric. Food Chem.* **1982**, *30*, 330–334. [CrossRef]

9.   Sancho, A.I.; Bartolomé, B.; Gómez-Cordovés, C.; Williamson, G.; Faulds, C.B. Release of ferulic acid from cereal residues by barley enzymatic extracts. *J. Cereal Sci.* **2001**, *34*, 173–179. [CrossRef]

10.   Mussatto, S.I.; Dragone, G.; Roberto, I.C. Ferulic and p-coumaric acids extraction by alkaline hydrolysis of brewer's spent grain. *Ind. Crop Prod.* **2007**, *25*, 231–237. [CrossRef]

11.   Bartolomé, B.; Faulds, C.B.; Williamson, G. Enzymic release of ferulic acid from barley spent grain. *J. Cereal Sci.* **1997**, *25*, 285–288. [CrossRef]

12.   Kumari, B.; Tiwari, B.K.; Hossain, M.B.; Brunton, N.P.; Rai, D.K. Recent advances on application of ultrasound and pulsed electric field technologies in the extraction of bioactives from agro-industrial by-products. *Food Bioprocess Technol.* **2018**, *11*, 223–241. [CrossRef]

13.   Naczk, M.; Shahidi, F. Phenolics in cereals, fruits and vegetables: Occurrence, extraction and analysis. *J. Pharm. Biomed. Anal.* **2006**, *41*, 1523–1542. [CrossRef] [PubMed]

14.   Moreira, M.M.; Morais, S.; Barros, A.A.; Delerue-Matos, C.; Guido, L.F. A novel application of microwave-assisted extraction of polyphenols from brewer's spent grain with HPLC-DAD-MS analysis. *Anal. Bioanal. Chem.* **2012**, *403*, 1019–1029. [CrossRef] [PubMed]

15.   Carciochi, R.A.; Sologubik, C.A.; Fernández, M.B.; Manrique, G.D.; D'Alessandro, L.G. Extraction of antioxidant phenolic compounds from brewer's spent grain: Optimization and kinetics modeling. *Antioxidants* **2018**, *7*, 45. [CrossRef] [PubMed]

16.   Meneses, N.G.T.; Martins, S.; Teixeira, J.A.; Mussatto, S.I. Influence of extraction solvents on the recovery of antioxidant phenolic compounds from brewer's spent grains. *Sep. Purif. Technol.* **2013**, *108*, 152–158. [CrossRef]

17.   Irakli, M.; Kleisiaris, F.; Kadoglidou, K.; Katsantonis, D. Optimizing extraction conditions of free and bound phenolic compounds from rice by-products and their antioxidant effects. *Foods* **2018**, *7*, 93. [CrossRef]

18.   Wang, J.; Sun, B.; Cao, Y.; Tian, Y.; Li, X. Optimisation of ultrasound-assisted extraction of phenolic compounds from wheat bran. *Food Chem.* **2008**, *106*, 804–810. [CrossRef]

19.   Fărcaş, A.; Tofana, M.; Socaci, S.; Scrob, S.; Salanta, L.-C.; Andrei, B. Preliminary study on antioxidant activity and polyphenols content in discharged waste from beer production. *J. Agroaliment. Process. Technol.* **2013**, *19*, 319–324.

20.   Piggott, C.O.; Connolly, A.; FitzGerald, R.J. Application of ultrafiltration in the study of phenolic isolates and melanoidins from pale and black brewers' spent grain. *Int. J. Food Sci. Technol.* **2014**, *49*, 2252–2259. [CrossRef]

21.   Spinelli, S.; Conte, A.; Del Nobile, M.A. Microencapsulation of extracted bioactive compounds from brewer's spent grain to enrich fish-burgers. *Food Bioprod. Process.* **2016**, *100*, 450–456. [CrossRef]

22.   Zuorro, A.; Iannone, A.; Lavecchia, R. Water–organic solvent extraction of phenolic antioxidants from brewers' spent grain. *Processes* **2019**, *7*, 126. [CrossRef]

23.   Fărcaş, A.C.; Socaci, S.A.; Dulf, F.V.; Tofană, M.; Mudura, E.; Diaconeasa, Z. Volatile profile, fatty acids composition and total phenolics content of brewers' spent grain by-product with potential use in the development of new functional foods. *J. Cereal Sci.* **2015**, *64*, 34–42. [CrossRef]

24. McCarthy, A.L.; O'Callaghan, Y.C.; Connolly, A.; Piggott, C.O.; FitzGerald, R.J.; O'Brien, N.M. Phenolic extracts of brewers' spent grain (BSG) as functional ingredients—Assessment of their DNA protective effect against oxidant-induced DNA single strand breaks in U937 cells. *Food Chem.* **2012**, *134*, 641–646. [CrossRef] [PubMed]

25. Moreira, M.M.; Morais, S.; Carvalho, D.O.; Barros, A.A.; Delerue-Matos, C.; Guido, L.F. Brewer's spent grain from different types of malt: Evaluation of the antioxidant activity and identification of the major phenolic compounds. *Food Res. Int.* **2013**, *54*, 382–388. [CrossRef]

26. Stefanello, F.S.; dos Santos, C.O.; Bochi, V.C.; Fruet, A.P.B.; Soquetta, M.B.; Dörr, A.C.; Nörnberg, J.L. Analysis of polyphenols in brewer's spent grain and its comparison with corn silage and cereal brans commonly used for animal nutrition. *Food Chem.* **2018**, *239*, 385–401. [CrossRef] [PubMed]

27. Escarpa, A.; González, M.C. Approach to the content of total extractable phenolic compounds from different food samples by comparison of chromatographic and spectrophotometric methods. *Anal. Chim. Acta* **2001**, *427*, 119–127. [CrossRef]

28. Sánchez-Rangel, J.C.; Benavides, J.; Heredia, J.B.; Cisneros-Zevallos, L.; Jacobo-Velázquez, D.A. The folin–ciocalteu assay revisited: Improvement of its specificity for total phenolic content determination. *Anal. Methods* **2013**, *5*, 5990–5999. [CrossRef]

29. Granato, D.; Shahidi, F.; Wrolstad, R.; Kilmartin, P.; Melton, L.D.; Hidalgo, F.J.; Miyashita, K.; Camp, J.V.; Alasalvar, C.; Ismail, A.B.; et al. Antioxidant activity, total phenolics and flavonoids contents: Should we ban in vitro screening methods? *Food Chem.* **2018**, *264*, 471–475. [CrossRef] [PubMed]

30. Harnly, J. Antioxidant methods. *J. Food Compos. Anal.* **2017**, *64*, 145–146. [CrossRef]

31. Singleton, V.L.; Orthofer, R.; Lamuela-Raventós, R.M. Analysis of total phenols and other oxidation substrates and antioxidants by means of Folin-Ciocalteu reagent. In *Methods in Enzymology*; Academic Press: San Diego, CA, USA, 1999; Volume 299, pp. 152–178.

32. Hossain, M.B.; Rai, D.K.; Brunton, N.P.; Martin-Diana, A.B.; Barry-Ryan, C. Characterization of phenolic composition in Lamiaceae spices by LC-ESI-MS/MS. *J. Agric. Food Chem.* **2010**, *58*, 10576–10581. [CrossRef]

33. Gangopadhyay, N.; Rai, D.K.; Brunton, N.P.; Gallagher, E.; Hossain, M.B. Antioxidant-guided isolation and mass spectrometric identification of the major polyphenols in barley (*Hordeum vulgare*) grain. *Food Chem.* **2016**, *210*, 212–220. [CrossRef]

34. Kumari, B.; Tiwari, B.K.; Walsh, D.; Griffin, T.P.; Islam, N.; Lyng, J.G.; Brunton, N.P.; Rai, D.K. Impact of pulsed electric field pre-treatment on nutritional and polyphenolic contents and bioactivities of light and dark brewer's spent grains. *Innov. Food Sci. Emerg. Technol.* **2019**, *54*, 200–210. [CrossRef]

35. Aura, A.-M.; Niemi, P.; Mattila, I.; Niemelä, K.; Smeds, A.; Tamminen, T.; Faulds, C.; Buchert, J.; Poutanen, K. Release of small phenolic compounds from brewer's spent grain and its lignin fractions by human intestinal microbiota in vitro. *J. Agric. Food Chem.* **2013**, *61*, 9744–9753. [CrossRef] [PubMed]

36. Friedman, M. Food browning and its prevention: An overview. *J. Agric. Food Chem.* **1996**, *44*, 631–653. [CrossRef]

37. Li, Y.; Li, S.; Lin, S.-J.; Zhang, J.-J.; Zhao, C.-N.; Li, H.-B. Microwave-assisted extraction of natural antioxidants from the exotic *Gordonia axillaris* fruit: Optimization and identification of phenolic compounds. *Molecules* **2017**, *22*, 1481. [CrossRef] [PubMed]

38. Bimakr, M.; Ganjloo, A.; Zarringhalami, S.; Ansarian, E. Ultrasound-assisted extraction of bioactive compounds from *Malva sylvestris* leaves and its comparison with agitated bed extraction technique. *Food Sci. Biotechnol.* **2017**, *26*, 1481–1490. [CrossRef] [PubMed]

39. Hernanz, D.; Nuñez, V.; Sancho, A.I.; Faulds, C.B.; Williamson, G.; Bartolomé, B.; Gómez-Cordovés, C. Hydroxycinnamic acids and ferulic acid dehydrodimers in barley and processed barley. *J. Agric. Food Chem.* **2001**, *49*, 4884–4888. [CrossRef] [PubMed]

40. Dobberstein, D.; Bunzel, M. Separation and detection of cell wall-bound ferulic acid dehydrodimers and dehydrotrimers in cereals and other plant materials by reversed phase high-performance liquid chromatography with ultraviolet detection. *J. Agric. Food Chem.* **2010**, *58*, 8927–8935. [CrossRef] [PubMed]

41. Faulds, C.; Sancho, A.; Bartolomé, B. Mono- and dimeric ferulic acid release from brewer's spent grain by fungal feruloyl esterases. *Appl. Microbiol. Biotechnol.* **2002**, *60*, 489–494.

42. Szwajgier, D.; Waśko, A.; Targoński, Z.; Niedźwiadek, M.; Bancarzewska, M. The use of a novel ferulic acid esterase from *Lactobacillus acidophilus* K1 for the release of phenolic acids from brewer's spent grain. *J. Inst. Brew.* **2010**, *116*, 293–303. [CrossRef]
43. Athanasios, M.; Georgios, L.; Michael, K. A rapid microwave-assisted derivatization process for the determination of phenolic acids in brewer's spent grains. *Food Chem.* **2007**, *102*, 606–611. [CrossRef]

 *antioxidants*

Article

Polyphenol-Rich Extracts Obtained from Winemaking Waste Streams as Natural Ingredients with Cosmeceutical Potential

Melanie S. Matos [1], Rut Romero-Díez [2], Ana Álvarez [2], M. R. Bronze [1,3,4],
Soraya Rodríguez-Rojo [2,*], Rafael B. Mato [2], M. J. Cocero [2] and Ana A. Matias [1,*]

[1]   Nutraceuticals & Bioactives Process Technology Group, Instituto de Biologia Experimental e
      Tecnológica (iBET), Av. República, Qta. Do Marquês, Estação Agronómica Nacional, Edifício iBET/ITQB,
      2780-157 Oeiras, Portugal
[2]   BioEcoUVa, Research Institute on Bioeconomy, High Pressure Processes Group, Department of Chemical
      Engineering and Environmental Technology, School of Engineering, University of Valladolid (UVa),
      Sede Mergelina Valladolid, 47011 Castilla y León, Spain
[3]   Instituto de Tecnologia Química e Biológica António Xavier (ITQB), Universidade Nova de Lisboa,
      Av. da República, 2780-157 Oeiras, Portugal
[4]   Faculty of Pharmacy, University of Lisbon (FFUL), Av. Prof. Gama Pinto, 1649-003 Lisbon, Portugal
*   Correspondence: sorayarr@iq.uva.es (S.R.-R.); amatias@ibet.pt (A.A.M.)

Received: 2 August 2019; Accepted: 23 August 2019; Published: 1 September 2019

Abstract: Phenolics present in grapes have been explored as cosmeceutical principles, due to their antioxidant activity and ability to inhibit enzymes relevant for skin ageing. The winemaking process generates large amounts of waste, and the recovery of bioactive compounds from residues and their further incorporation in cosmetics represents a promising market opportunity for wine producers and may contribute to a sustainable development of the sector. The extracts obtained from grape marc and wine lees, using solid–liquid (SL) extraction with and without microwave (MW) pretreatment of the raw material, were characterized in terms of antioxidant activity through chemical (ORAC/HOSC/HORAC) and cell-based (keratinocytes—HaCaT; fibroblasts—HFF) assays. Furthermore, their inhibitory capacity towards specific enzymes involved in skin ageing (elastase; MMP-1; tyrosinase) was evaluated. The total phenolic and anthocyanin contents were determined by colorimetric assays, and HPLC–DAD–MS/MS was performed to identify the main compounds. The MW pretreatment prior to conventional SL extraction led to overall better outcomes. The red wine lees extracts presented the highest phenolic content (3 to 6-fold higher than grape marc extracts) and exhibited the highest antioxidant capacity, being also the most effective inhibitors of elastase, MMP-1 and tyrosinase. The results support that winemaking waste streams are valuable sources of natural ingredients with the potential for cosmeceutical applications.

Keywords: phenolics; antioxidants; anti-ageing; skin whitening; grape marc; wine lees

---

1. Introduction

Europe is responsible for the largest share of wine production globally, accounting for more than 60% of the world's entire production [1]. The winemaking process generates large amounts of solid organic waste and by-products, including grape marc (62%), wine lees (14%), grape stalk (12%), and dewatered sludge (12%) [2]. It is estimated that 14.5 million tons of byproducts from wineries are generated annually in Europe alone [3], and the discarding of these dregs may potentially cause environmental issues, due to a low pH and the presence of phytotoxic and antibacterial phenolic substances resisting biological degradation [4]. Grape and wine (poly)phenols are already exploited as cosmetic ingredients due to their renowned antioxidant activities. However, the exploitation of winery wastes is not common yet. In this regard, and since waste streams from the vinification process may present an environmental hazard, the recovery of high-added value bioactive compounds, such as (poly)phenols, from winemaking residues seems a promising market opportunity for wine producers and may contribute to a sustainable development of the sector.

Aged skin is known to have a compromised barrier function, resulting in a dry appearance and susceptibility to environmental aggressors, and therefore is an enhanced risk for skin disorders [5]. Apart from the uneven pigmentation of the epidermis and due to alterations in tyrosinase activity amongst melanocytes [6], the main changes in ageing skin occur at the level of the dermal connective tissue and are essentially translated into the loss of mature collagen and alterations in the elastic network [7]. Among extracellular matrix (ECM)-degrading enzymes, matrix metalloproteinases (MMPs) and elastolytic enzymes (elastases) can be found. These endopeptidases are responsible for the turnover of several ECM components, including all the types of collagen and elastin, playing important roles in numerous physiological processes, such as tissue repair and remodeling, cell migration and differentiation, or wound healing [8–10]. However, an exacerbated amount of these enzymes in their active form is the driving cause of several pathological conditions, including accelerated skin ageing.

Reactive oxygen species (ROS), resulting from electron leakage during aerobic metabolism and upon exposure to environmental factors, are unstable species capable of inducing damage to several biomolecules, leading to altered functionality. Hence, to counteract their effect, there are natural antioxidant defenses in the organism with the function of maintaining ROS within physiologically acceptable levels. However, a fraction of the formed ROS recurrently evades this antioxidant control [11]. These oxidant species severely contribute to the skin ageing process, either through direct damage to biomolecules therein, or through interference with signaling pathways within keratinocytes and fibroblasts, thus altering the expression balance of MMPs, procollagen and pro-inflammatory cytokine genes [12,13].

It is known that phenolic compounds, including anthocyanins, which are the major phenolics present in red grapes, not only possess renowned antioxidant properties, but also have the ability to directly inhibit enzymes enrolled in the skin ageing process, namely tyrosinase, collagenase (MMP-1) and elastase [10,14]. In this regard, it is expected that these bioactive compounds will also be encountered in winery residues, such as grape marc, red wine lees and Port wine lees. Grape marc consists of a pressed mixture of grape pulp, skins and seeds, obtained from the separation of the solid fraction of the must from its liquid fraction. Wine lees are the deposit obtained after the fermentation of wine, mainly composed of dead yeast and bacteria, tartaric salts, precipitated tannins, organic and inorganic matter, and free phenolic compounds [15]. The main difference between the vinification process of red wine and Port wine is that, in the latter, alcoholic fermentation is interrupted due to the fortification with wine spirits, which explains the sweetness of some wines and also the high alcoholic grade of Port wine [16].

The work presented herein was focused on the comparative assessment of the cosmeceutical potential of natural extracts obtained from three different winemaking waste streams: Wine lees resulting from alcoholic fermentation of red wine, port wine lees, and red grape marc. For this purpose, all the extracts were characterized in terms of antioxidant activity and inhibitory capacity towards tyrosinase, elastase and MMP-1 (collagenase), through chemical, enzymatic and cell-based assays. Additionally, the phytochemical characterization was carried out by colorimetric assays as well as HPLC-DAD-MS/MS. The main goal of this study was to validate the antioxidant, anti-ageing and skin whitening potential of bioactive compounds recovered from winery byproducts, for their application in the cosmetic industry.

2. Materials and Methods

2.1. Chemicals

The chemicals used for the determination of the phenolic content and anthocyanin content were gallic acid (Fluka, Steinheim, Germany), sodium carbonate and *Folin-Ciocalteu* reagent (Panreac, Barcelona, Spain), potassium chloride (Sigma-Aldrich, St. Louis, MO, USA), and sodium acetate trihydrate (Sigma-Aldrich, Steinheim, Germany). For HPLC analyses, acetonitrile (Panreac, Barcelona,

Spain) and formic acid (VWR-CHEM, Radnor, PA, USA) were used. For the antioxidant activity assays, the reagents were disodium fluorescein, 2,2′-azobis(2-methylpropionamidine)dihydrochloride (AAPH), cobalt (II) fluoride tetrahydrate (Sigma-Aldrich, St. Louis, MO, USA), (+/−)-6-hydroxy-2,5,7,8-tetramethylchroman-2-carboxylic acid (Trolox) (Aldrich, St. Louis, MO, USA), caffeic acid (Sigma-Aldrich, St. Louis, MO, USA), hydrogen peroxide 30 wt. % in water and iron (III) chloride (Sigma-Aldrich, Steinheim, Germany), and acetone (Sigma-Aldrich, St. Louis, MO, USA). For the preparation of phosphate-buffered saline (PBS) 75 mM pH 7.40, potassium phosphate monobasic anhydrous (KH_2PO_4) (Amresco, Solon, OH, USA), sodium phosphate dibasic dihydrate ($Na_2HPO_4 \cdot 2H_2O$) (Sigma-Aldrich, Steinheim, Germany), potassium chloride (KCl) and sodium chloride (NaCl) (Sigma-Aldrich, St. Louis, MO, USA) were used. Sodium phosphate dibasic dihydrate ($Na_2HPO_4 \cdot 2H_2O$) and sodium phosphate monobasic monohydrate ($NaH_2PO_4 \cdot H_2O$) (Sigma-Aldrich, Steinheim, Germany) were used to prepare sodium phosphate buffer solution (SPB) 75 mM pH 7.40. The tyrosinase inhibition was assessed using 3,4-dihydroxy-L-phenylalanine (L-DOPA) (Sigma-Aldrich, St. Louis, MO, USA), mushroom tyrosinase (Sigma-Aldrich, St. Louis, MO, USA), and kojic acid (Sigma-Aldrich, St. Louis, MO, USA). The elastase inhibition was assayed with porcine pancreatic elastase (PPE) type III and N-succinyl-Ala-Ala-Ala-*p*-nitroanilide (AAAPVN) (Sigma-Aldrich, St. Louis, MO, USA); for MMP-1 inhibition reagents were recombinant (expressed in *E. coli*) MMP-1 (Sigma-Aldrich, St. Louis, MO, USA) and MMP fluorogenic substrate (Enzo Life Sciences, Farmingdale, NY, USA). Tyrosinase assay buffer (SPB 0.1 M, pH 6.8) was prepared with sodium phosphate dibasic dihydrate and sodium phosphate monobasic monohydrate (Sigma-Aldrich, Steinheim, Germany); Tris base (Sigma-Aldrich, St. Louis, MO, USA) and hydrochloric acid (HCl) 37% (*w/w*) (Honeywell Riedel-de-Häen, Hanover, Germany) were used to prepare elastase assay buffer (Tris-HCl 0.1 M, pH 8); the buffer used in the MMP-1 assay (0.05 M Tris-HCl, pH 7.5) was prepared with Tris-HCl (Fluka, Steinheim, Germany), calcium chloride dihydrate ($CaCl_2 \cdot 2H_2O$) (Riedel-de Haën, Seelze, Germany), sodium azide (NaN_3) (Sigma-Aldrich, Steinheim, Germany), Brij 35 (Fisher Scientific, Geel, Belgium), zinc sulfate heptahydrate ($ZnSO_4 \cdot 7H_2O$) (Merck, Darmstadt, Germany), and sodium chloride (Sigma-Aldrich, St. Louis, MO, USA). For cell-based assays, the high glucose Dulbecco's modified eagle medium (DMEM) (Gibco – Thermo Fisher Scientific, Grand Island, NY, USA) and Iscove's Modified Dulbecco's Medium (IMDM – GlutaMAX™) (Gibco – Thermo Fisher Scientific, Paisley, UK) were used for cell culturing, and the cells were subcultured with 0.25% trypsin-EDTA (Gibco – Thermo Fisher Scientific, Paisley, UK). Both DMEM and IMDM were supplemented with fetal bovine serum (FBS) (Biowest, Nuaillé, France), and Penicillin-Streptomycin (Gibco – Thermo Fisher Scientific, Grand Island, NY, USA). CellTiter 96® AQueous One Solution Cell Proliferation Assay (MTS/5-(3-carboxymethoxyphenyl)-2-(4,5-dimethylthiazoly)-3-(4-sulfophenyl)tetrazolium, inner salt), from Promega (Madison, WI, USA), was used for cytotoxicity evaluation. For the cellular antioxidant activity assays, 2′,7′-dichlorofluorescin diacetate (DCFH-DA) (Sigma-Aldrich, St. Louis, MO, USA) and tert-butyl hydroperoxide (TBHP) 70% wt. in water (Sigma-Aldrich, Steinheim, Germany) were used.

2.2. Samples

The red wine lees from alcoholic fermentation and grape marc of *Tempranillo* grapes from Ribera del Duero Denomination of Origin were kindly provided by Matarromera winery (Valladolid, Spain) in 2015. The port wine lees were generously provided by Sogrape Vinhos S.A. (Porto, Portugal) in 2015.

The extracts from the three different winemaking waste streams were prepared by conventional solid-liquid (SL) extraction, preceded or not by microwave (MW) pretreatment. The grape marc extraction procedure was followed according to a process intensification study described by Álvarez [17], and the optimized extraction parameters for the port and red wine lees were used as designated by Romero-Díez [15]. In brief, the SL extraction of grape marc was carried out with 50:50 (% *v/v*) EtOH:H_2O (water acidified to pH 1 with sulfuric acid) at 60 °C. For MW-pretreated grape marc, the SL extraction was performed at 60 °C with the same extraction solvent after MW irradiation of the samples to achieve a maximum temperature of 80 °C (60 s). The SL extractions from wine lees were performed at 25 °C with 50:50 (% *v/v*) EtOH:H$_2$O (water acidified to pH 2.5 with hydrochloric acid). For MW-pretreated wine lees, the sample (with a solvent mixture of 60:40 (% *v/v*) EtOH:H$_2$O) was irradiated for 90 s to reach a temperature of 115 °C; immediately after, the mixture was cooled down and additional solvent was added to carry out the extraction at 25 °C as described above. All chosen extraction temperatures resulted from an optimization process aiming at maximizing the extraction of anthocyanins from grape marc and wine lees, which are the main phenolics present in these raw materials. These optimization studies [15,17,18] took into consideration the compromise between the thermal effect that enhances the extraction yield of phenolic compounds while reducing the extraction time, and the vulnerability of phenolics to thermal degradation. In the case of MW-pretreatment, it was concluded that the high temperatures achieved after irradiation did not lead to degradation of phenolics due to the short duration of the heating. The MW pretreatments were carried out in a CEM Discover Microwave (CEM Corporation, Matthews, NC, USA) using a maximum power of 300 W. Ethanol was eliminated from the samples by rotary vacuum evaporation until only 5% ethanol was achieved, and phytochemical as well as antioxidant activity characterization was performed. For the enzymatic and cell-based assays, the extracts were dried in a CentriVap Concentrator (Labconco, Kansas City, MO, USA), solubilized in DMSO and then stored at −20 °C.

2.3. Methods

2.3.1. Phytochemical Characterization

Total Phenolic Content (TPC)

The total phenolic content of the extracts was determined by the Folin-Ciocalteu (FC) method, relying on the electron transfer from the phenolic compounds to phosphomolybdic/phosphotungstic acid complexes in alkaline medium. This resulted in the formation of blue complexes which absorbance at 765 nm is proportional to the amount of phenolics [19]. This assay was based on previous work [20], by adding FC reagent and a saturated sodium carbonate solution to the samples, and adapted for a Spark 10M (Tecan Group Ltd., Zürich, Switzerland) spectrophotometer microplate reader. The absorbances were measured against the blank and the TPC values were calculated from a gallic acid standard curve and expressed as milligrams of gallic acid equivalents (GAE) per gram of dry extract.

Total Anthocyanin Content (TAC)

The total monomeric anthocyanin pigment content was assessed by a pH differential method, following the protocol described in the AOAC Official Method 2005.02 [21]. This method relies on the color change of monomeric anthocyanin pigments depending on the pH, and their differential absorbance at 520 nm. The readings were carried out in a Spark 10M (Tecan Group Ltd., Zürich, Switzerland) spectrophotometer microplate reader, and the results were calculated as milligrams of malvidin-3-O-glucoside equivalents per gram of dry extract, since this anthocyanin is a major compound in all the extracts.

High Performance Liquid Chromatography–Mass Spectrometry (HPLC-DAD-MS/MS)

The samples were analyzed by HPLC-DAD-MS/MS, using a Waters Alliance 2695 Separation Module (Waters, Ireland) system equipped with a quaternary pump, a degasser, an autosampler and a column oven. The liquid chromatography system was coupled to a photodiode array detector 996 PDA, and to a mass spectrometer MicroMass Quattromicro® API (Waters, Ireland). All data were acquired and processed by MassLynx® 4.1 software.

The chromatographic separation of compounds was carried out in a reversed-phase LiChrospher® 100 RP-18 5μm LiChroCART® 250-4 column inside a thermostatic oven at 35 °C. A binary mobile phase was used, at a flowrate of 0.3 mL/min, with eluent A composed of formic acid (0.5% *v/v* in ultrapure water) and eluent B of acetonitrile. The gradient program used was 99:1 A:B for 5 min, from 99:1 A:B to 40:60 A:B in 40 min, from 40:60 A:B to 10:90 A:B in 45 min, held isocratically (90% B) for 10 min, from 10:90 A:B to 99:1 A:B in 10 min, and finally held isocratically (99:1 A:B) for 10 min. An injection volume of 20 µL was used. The absorption spectra were acquired from 210 to 600 nm by a photodiode array detector. Mass spectrometry was performed using an electrospray ion source in the negative and positive ion mode, with the temperature set at 120 °C, the capillary voltage at 2.5 kV, and the source voltage at 30 V. The compounds separated by HPLC were ionized and the mass spectra were recorded in a full scan mode, with m/z range between 100 and 1500. High purity nitrogen was used as drying and nebulizing gas, and ultrahigh purity argon was used as collision gas.

2.3.2. Antioxidant Activity

The antioxidant activity of the extracts was assessed towards different ROS, through three complementary antioxidant assays: Oxygen radical absorbance capacity (ORAC), hydroxyl radical scavenging capacity (HOSC), and hydroxyl radical averting capacity (HORAC). All three assays rely on the capacity of the samples to prevent the oxidation of disodium fluorescein (FL). In all cases, fluorescence (Ex/Em 485 ± 20/528 ± 20 nm) emitted by the reduced form of FL was recorded over time, at 37 °C, in a FL800 microplate fluorescence reader (Bio-Tek Instruments, Winooski, VT, USA), under the control of Gen5 software.

The ORAC evaluates the antioxidant capacity of the tested samples towards peroxyl radicals ($ROO^\bullet$) generated during thermal decomposition of AAPH. This assay was based in the method developed by Huang [22], with some modifications. Briefly, FL was added to sample dilutions and the resulting mixture was equilibrated to 37 °C, then, the reaction was initiated by the addition of AAPH and fluorescence was recorded for 40 min. Final concentrations in the reaction mixture were 2.25×10^{-4} mM FL and 19.12 mM AAPH, and all solutions were prepared in a phosphate-buffered saline (PBS), 75 mM, pH 7.4.

The HOSC estimates the capacity of the samples to scavenge hydroxyl radicals ($^\bullet OH$) generated from a Fe(III)-driven Fenton-like reaction. The assay was performed as described by Moore [23], with slight modifications. Briefly, FL, hydrogen peroxide (H_2O_2) and iron (III) chloride ($FeCl_3$) were added to sample dilutions, and fluorescence was measured for 60 min. The final concentrations of the reagents were 5.64×10^{-5} mM FL, 26.67 mM H_2O_2 and 0.68 mM $FeCl_3$. FL solution was prepared in a sodium phosphate buffer (SPB), 75 mM, pH 7.4, $FeCl_3$ and H_2O_2 solutions were prepared in MilliQ water, while sample dilutions were made in Acetone:MilliQ water 50% (*v/v*).

The HORAC aims to evaluate the capacity of a given sample to prevent the generation of hydroxyl radicals ($^\bullet OH$) by a Co(II)-mediated Fenton-like reaction. The procedure was performed based on the method described by Ou [24], with some modifications. In brief, FL, hydrogen peroxide (H_2O_2) and cobalt (II) fluoride (CoF_2) were added to the sample dilutions, and fluorescence was measured for 60 min. The final concentrations were 5.64×10^{-5} mM FL, 26.67 mM H_2O_2 and 0.41 mM CoF_2. The FL solution was prepared in a sodium phosphate buffer (SPB), 75 mM, pH 7.4, CoF_2 and H_2O_2 solutions were prepared in MilliQ water, while the sample dilutions were made in Acetone:MilliQ water 50% (*v/v*).

The ORAC and HOSC values were calculated from a trolox standard curve and expressed as micromoles of Trolox equivalents (TE) per gram of dry extract, whereas for HORAC, caffeic acid was used as the standard and the results were expressed as micromoles of caffeic acid equivalents (CAE) per gram of dry extract. In all three assays, the calculations took into consideration the dilution effect on the antioxidant capacity [25].

2.3.3. Enzymatic Assays

Inhibition of Tyrosinase

The tyrosinase inhibitory capacity of the extracts was determined spectrophotometrically, using mushroom tyrosinase and L-DOPA as the substrate, as reported in the literature [26]. Tyrosinase converts L-DOPA to Dopaquinone, which in turn cyclizes to form Dopachrome. The dopachrome formation can be monitored by measuring the absorbance at 475 nm. Shortly after, L-DOPA was added to tyrosinase in the presence of the sample dilutions, to a final concentration of 6 U/mL tyrosinase and 0.5 mM L-DOPA. After a 30 min incubation at 37 °C, absorbance was measured at 475 nm in a Spark 10M (Tecan Group Ltd., Männedorf, Zürich, Switzerland). All reagents were prepared in SPB, 0.1 M, pH 6.8. The calculations were made as follows:

$$\% \ inhibition = \frac{(A control - A sample)}{A control} * 100 \tag{1}$$

where $A_{control}$ and A_{sample} stand for the A_{475} in the absence or presence of the sample, respectively. The inhibitory potential of the extracts was evaluated with increasing concentrations, in order to establish dose-dependent relationships and determine the half maximal inhibitory concentration (IC_{50}) values, meaning the capacity of the samples to inhibit the enzymatic activity to an extent of 50%.

Inhibition of Elastase

The elastase inhibitory capacity of the extracts was determined by a colorimetric assay, using porcine pancreatic elastase (PPE) and N-succinyl-Ala-Ala-Ala-p-nitroanilide (AAAPVN) as the substrate. Notably, *p*-nitroaniline is formed after cleavage of the substrate and its formation can be monitored by measuring the absorbance at 410 nm. Briefly, elastase was added to the sample dilutions, and after equilibrating the temperature to 25 °C for 20 min, the reaction was initiated by the addition of the substrate. After a 20 min incubation, absorbance was measured at 410 nm in a Spark 10M (Tecan Group Ltd., Männedorf, Zürich, Switzerland). The procedure was carried out in a Tris-HCl buffer (0.1 M, pH 8), and the final concentrations were 0.03 U/mL elastase and 0.05 mg/mL AAAPVN. The calculations were made as described in Equation (1), and several concentrations of the extracts were tested in order to determine the IC_{50} values.

Inhibition of MMP-1

The capacity of the extracts to inhibit matrix metalloproteinase-1 (MMP-1) was assayed using human recombinant MMP-1, and a fluorogenic peptide as MMP substrate, displaying strong fluorescence (Ex/Em 340/440 nm) once cleaved by the enzyme. Briefly, the fluorogenic substrate was added to MMP-1 in the presence of the extract dilutions, with the final concentrations of 0.2 µg/mL MMP-1 and 1 µM fluorogenic substrate. The reaction was allowed to occur for 20h at 37 °C, and then fluorescence was measured in a multimode microplate reader (Spark 10M, Tecan Group Ltd., Männedorf, Zürich, Switzerland). Tris-HCl 50 mM, pH 7.5, with 10 mM $CaCl_2$, 150 mM NaCl, 0.02% (*w/v*) NaN_3, 0.05% (*w/v*) Brij 35 and 0.05 mM $ZnSO_4$ was used as the assay buffer, according to the literature [27]. The calculations were made as described in Equation (1), and several concentrations of the extracts were tested in order to determine the IC_{50} values.

2.3.4. Cell-based Assays

Cell Culture

The human immortalized non-tumorigenic keratinocyte cell line HaCaT (CLS, Germany) was cultured in high glucose, high pyruvate, Dulbecco's modified eagle medium (DMEM), whereas the human foreskin fibroblasts (HFF) cell line CCD-1112Sk (ATCC, USA) was cultured with Glutamax™ Iscove's Modified Dulbecco's Medium (IMDM). Both culture media were supplemented with 10% (*v/v*) heat-inactivated fetal bovine serum (FBS), 100 units/mL penicillin and 100 µg/mL streptomycin. All experiments were performed in culture media supplemented with only 0.5% FBS and no antibiotic. For every assay, the cells were seeded in 96-well TC (tissue culture)-treated microplates at a density of 1.4×10^5 cells/cm² (HaCaT) or 3.1×10^4 cells/cm² (HFF) and allowed to reach confluence. The cells were cultured in a humidified atmosphere at 37 °C with 5% CO_2.

Cytotoxicity Evaluation

In order to determine the nontoxic concentrations of the extracts for further studies, the cells were exposed to several concentrations of the extracts diluted in culture medium for 24 h. The well content was then removed, and the cells were washed twice with PBS. A solution of 1.6% *v/v* MTS in the medium was added to the cells for 3 h, and absorbance was measured at 490 nm in a multimode microplate reader (Spark 10M, Tecan Group Ltd., Männedorf, Zürich, Switzerland). Cell viability was determined as a percentage of control, after blank subtraction. The MTS assay is based on the reduction of a tetrazolium salt by viable cells to generate a colored, aqueous soluble formazan product, of which absorbance can be measured at 490 nm. The amount of formazan produced is directly proportional to the number of viable cells.

Cellular Antioxidant Activity

The capacity of the extracts to inhibit ROS production in the cells was evaluated using two different approaches: Pre-incubation and co-incubation. In both cases, 2′,7′-dichlorofluorescin diacetate (DCFH-DA) was used as a fluorescent probe. Non-fluorescent DCFH-DA readily diffuses through the cell membrane and once in the intracellular medium, the diacetate moiety is cleaved by cellular esterases giving rise to the more polar 2′,7′-dichlorodidhydrofluorescein ($DCFH_2$) which remains trapped within the cell. ROS from intrinsic oxidative stress or generated by an oxidative stress inducer easily diffuse into the cell, where they oxidize $DCFH_2$ to its fluorescent form, 2′,7′-dichlorofluorescein (DCF). The accumulation of DCF in the cells may be measured by an increase in fluorescence (Ex/Em 485/528 nm), which is proportional to the amount of ROS [28].

In the pre-incubation approach, cells were treated with selected non-toxic concentrations of the samples for 1 h or 24 h (cytotoxicity data in Supplementary Material, Figures S1 and S2), and then incubated with 25 µM DCFH-DA in PBS for 1 h. Fluorescence was measured at this point in order to assess the antioxidant effect of the samples towards intrinsic ROS. DCFH-DA was then removed and a non-cytotoxic concentration of *tert*-butyl hydroperoxide (TBHP) was added to the cells in PBS (0.625 mM for HaCaT and 1.25 mM for HFF). After 1 h, fluorescence was measured. In the co-incubation approach, the cells were incubated with 25 µM DCFH-DA for 1 h in PBS, and then the chosen concentrations of the stress inducer and extract were simultaneously added to the cells, in PBS. After 1 h, fluorescence was measured. All the results were presented as fluorescence percentages relative to the untreated control. Fluorescence measurements were performed in a FL800 microplate fluorescence reader (Bio-Tek Instruments, Winooski, VT, USA), with fluorescence filters (Ex/Em 485 ± 20/528 ± 20 nm).

Protection against Oxidant-Induced Cytotoxicity

To assess the potential of the samples to prevent TBHP-induced cytotoxicity, the cells were incubated with selected non-toxic concentrations of the extracts for 24 h. After this period, the cells were incubated for 1 h with a concentration of TBHP capable of inducing cytotoxicity (20 mM for HaCaT and 10 mM for HFF), and then cell viability was measured using the MTS assay. The absorbance measurements were performed in an EPOCH 2 microplate reader (Bio-Tek Instruments, Winooski, VT, USA).

2.3.5. Statistical Analysis

All the results are expressed as the mean ± standard deviation (SD), obtained from at least three independent experiments. Statistical analysis of the results was performed using GraphPad Prism 6 software (GraphPad Software, Inc., La Jolla, CA, USA). When homogeneous variance was confirmed, the results were analyzed by one-way analysis of variance (ANOVA), followed by the Tukey test for multiple comparisons. In the case of heterogeneous variances, an appropriate unpaired student *t*-test was performed in order to determine whether the means were significantly different. A *p*-value ≤ 0.05 was accepted as statistically significant in all cases.

3. Results and Discussion

3.1. Phytochemical and Antioxidant Activity Characterization

3.1.1. Total Phenolic Content (TPC) and Total Anthocyanin Content (TAC)

In a first approach, this study aimed to assess the effect of MW-pretreatment in the extraction methodology, by characterizing the extracts in terms of total phenolic content (TPC). Moreover, since the extracts studied in this work were obtained from red grape marc and red wine lees, and given anthocyanins are a relevant class of compounds encountered in these matrices, total anthocyanin content (TAC) was also evaluated.

Microwaves (MW) have been widely used to assist the extraction of several compounds from plant matrices, as they generally result in higher extraction yields, shorter extraction times, and reduced amounts of solvent needed, when compared to conventional extraction techniques. This is because water molecules present in the matrix absorb MW energy, leading to rapid heating and evaporation of intracellular water, which in turn causes disruption of the plant cell membrane-limited compartments, improving the mass transfer process of substances of interest from the raw material to the extraction solvent [29,30]. To overcome the scale-up limitations of a full-time low frequency microwave-assisted extraction (MAE), particularly the non-uniform irradiation of a large vessel, MW-pretreatment has been suggested, which comprises a short time irradiation with a higher MW frequency that allows the material to be homogeneously irradiated. The short duration (<120 s) peak of energy obtained in the MW-pretreatment is proposed to avoid degradation of the active compounds while maintaining the MW thermal effect that accelerates the extraction [17].

In both the Port and red wine lees extracts, the higher values of TPC and TAC are observed in the case of MW-pretreated matrices (Table 1), implying that MW treatment prior to SL extraction positively influences (poly)phenol and anthocyanin richness. This finding agrees with the abovementioned principle of MW-pretreatment, and equivalent results have been reported in the literature for the same matrices [15]. It was found that TAC in MW80 GM (2.7 mg malv-3-O-gl/g extract) was significantly higher (>1.5-fold increase) than GM (1.7 mg malv-3-O-gl/g extract). On the other hand, the grape marc conventional extract (GM) presents a higher TPC (83.9 mg GAE/g extract) than the grape marc extract obtained after MW-pretreatment (45.9 mg GAE/g extract). These results suggest that the energy of the MW-pretreatment used may have not been enough to improve the extraction of all subclasses of phenolics, but instead selectively increased anthocyanin richness. When comparing our results with the ones obtained by Álvarez [17], it is evident that the order of magnitude for both TPC and TAC of grape marc extracts studied herein is much lower. This may be explained by the usage of different batches of grape marc (2014 versus 2015 vintage) and thus the different composition of the raw material used, since crops from different years may present considerable variability in sugar content and nutritional composition. Nevertheless, the trend observed for TAC agrees with the one found for grape marc in the literature [17].

Still, the red wine lees extracts, both with and without MW pretreatment (MW RW and RW, respectively), are the richest of the studied extracts in terms of both TPC and TAC. In particular, the TPC values of the red wine lees extracts are at least two-fold higher than those of the Port wine lees extracts, and three-fold higher than those of the grape marc extracts, whereas the TAC values of the red wine lees extracts are at least two-fold higher than those of the Port wine lees extracts and ten-fold higher than those of the grape marc extracts. These results may be due to the sugars found in the Port wine and grape marc, that are extracted along with phenolics, leading to a reduced (poly)phenol richness per mass of extract when compared to the red wine lees extracts. Additionally, the grape marc extracts have significantly lower TAC values than both wine lees extracts, probably because of the higher availability of anthocyanins in wine lees due to the extraction of phenolics from grape skin taking place in the winemaking process [31]. Moreover, it is known that the capacity of wine lees to adsorb colorants leads to a concentration of anthocyanins up to ten times higher than in red grape skin [32].

Table 1. Phytochemical composition and antioxidant activity of the extracts. The results identified with different letters (a to f) in the same column are statistically different (p-value $\leq$ 0.05).

	Extract	Phytochemical Composition		Antioxidant Activity		
		TPC (mg GAE/g Extract)	TAC (mg malv-3-O-gl/g Extract)	ORAC (µmol TE/g Extract)	HOSC (µmol TE/g Extract)	HORAC (µmol CAE/g Extract)
Wine lees	RW	237.4 ± 7.7 [a]	28.6 ± 2.4 [a]	3167 ± 189 [a]	3680 ± 163 [a]	1932 ± 130 [a]
	MW RW	266.0 ± 5.6 [b]	29.5 ± 2.3 [a]	3500 ± 223 [a]	4776 ± 268 [b]	2625 ± 135 [b]
	P	64.0 ± 2.7 [c]	6.1 ± 0.7 [b]	451 ± 26 [b]	837 ± 49 [c]	458 ± 29 [c]
	MW P	114.5 ± 4.7 [d]	11.5 ± 1.0 [c]	716 ± 41 [c]	1285 ± 95 [d]	776 ± 49 [d]
Grape marc	GM	83.9 ± 2.0 [e]	1.7 ± 0.1 [d]	481 ± 30 [b]	746 ± 49 [c]	305 ± 28 [e]
	MW80 GM	45.9 ± 1.5 [f]	2.7 ± 0.3 [e]	448 ± 31 [b]	441 ± 34 [e]	198 ± 19 [f]

RW—red wine lees conventional extract; MW RW—MW-pretreated red wine lees extract; P—Port wine lees conventional extract; MW P—MW-pretreated Port wine lees extract; GM—grape marc conventional extract; MW80 GM—MW-pretreated grape marc extract (max. temp. 80 °C).

3.1.2. Identification of Compounds by HPLC-MS/MS Analysis

An HPLC apparatus coupled to a diode array detector (DAD) and a mass spectrometer (MS) was used to identify the main compounds present in the three raw materials used. The identification was carried out taking into consideration the absorption of compounds at four wavelengths (280, 320, 360 and 520 nm, for phenolics in general, phenolic acids, flavonols, and anthocyanins, respectively), the *m/z* peaks corresponding to precursor and daughter ions, a comparison of chromatographic profiles with those of standard compounds, databanks [33,34], and studies already reported in the literature for comparable matrices [15,35–40].

The TPC and TAC obtained by this method were consistent with the results obtained by the colorimetric assays, in terms of relative quantification. The relative flavonol amounts in the extracts were also estimated by calculating the total peak area of the chromatograms at 360 nm, and it was found that the order of flavonol content was: RW > MW RW > MW P > P > MW80 GM > GM.

In all cases, the extracts obtained from the same matrix had the same qualitative composition, differing only in quantitative composition, depending on the extraction methodology. An example is presented in Figure 1 for RW and MW RW, corroborating the positive effect of MW-pretreatment on the amount of phenolics extracted. Since the chromatographic profiles of the extracts obtained from the same waste stream matrices are qualitatively identical, the identifications presented in Table 2 are organized by raw materials, along with m/z values, fragment ions and phenolic subclasses. The qualitative composition of the different raw materials is similar, however, some differences can be pointed out. In particular, some anthocyanin conjugates (petunidin-, malvidin-, and peonidin-3-O-6″-*p*-acetylglucosides), that were identified in the Port and red wine lees, were not detected in grape marc. Once again, this may be explained by the higher availability of phenolics in wine lees rather than in grape marc. However, two pyranoanthocyanins (vitisin A and 10-carboxypyranomalvidin-3-6″-*p*-coumaroylglucoside), resulting from the reactions between malvidin-3-O-glucoside and pyruvic acid, were identified in the three raw materials. These anthocyanin-derived compounds can only be formed after alcoholic fermentation occurs, as pyruvic acid is a product of this reaction [38]. Since grape marc is only separated after the alcoholic fermentation is initiated, it is plausible that these pigments are found, not only in wine lees, but also in grape marc. These findings are corroborated by literature, since vitisin A has already been identified in Port and red wine lees [15], and several pyranoanthocyanins, including vitisin A and 10-carboxypyranomalvidin-3-6″-*p*-coumaroylglucoside, have been detected in red grape marc [41].

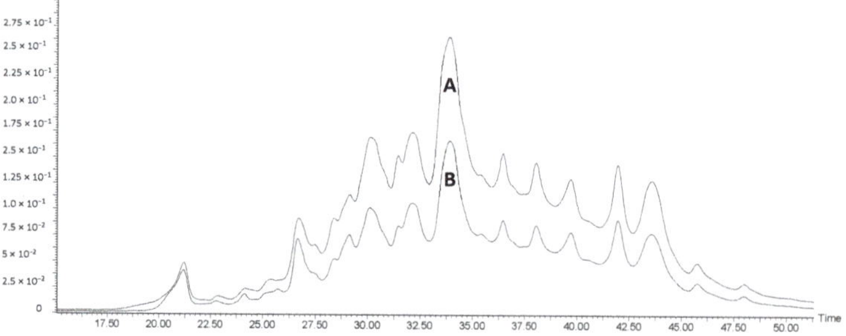

Figure 1. Chromatograms at 280 nm of MW RW (**A**) and RW (**B**), as obtained by HPLC.

Table 2. The putative identification of phenolic compounds present in grape marc, red wine lees and Port wine lees, as obtained by HPLC-DAD-MS/MS. The retention times, *m/z* values and respective fragments, as well as the phenolic subclass, are presented.

Retention Time (min)	*m/z*	Ionic Species	Fragment Ions	Putative Identification	Phenolic Subclass	Grape Marc	Red Wine Lees	Port Wine Lees
23.2	169	[M–H]⁻	125	Gallic acid	Phenolic acid	✓	✓	✓
26.6	616	[M–H]⁻	466, 307, 272,167, 134	2-S-glutathionylcaftaric acid	Phenolic acid	✓	✓	✓
27.7	577	[M–H]⁻	289	Procyanidin dimer	Flavanol	✓	✓	✓
28.4	311	[M–H]⁻	179, 149, 135	Caftaric acid	Phenolic acid	✓	✓	✓
28.9	865	[M–H]⁻	577, 289	Procyanidin trimer	Flavanol	✓	✓	✓
29.8	289	[M–H]⁻	245	Catechin	Flavanol	✓	✓	✓
30.1	465	[M–H]⁺	303	Delphinidin-3-O-glucoside	Anthocyanin	✓	✓	✓
30.2	577	[M–H]⁻	289, 175, 129	Procyanidin dimer	Flavanol	✓	✓	✓
30.2	463	[M–H]⁻	300	Quercetin-3-O-glucoside	Flavonol	✓	✓	✓
31.1	295	[M–H]⁻	163, 149, 119	Coutaric acid	Phenolic acid	✓	✓	✓
31.5	289	[M–H]⁻	245	Epicatechin	Flavanol	✓	✓	✓
31.9	449	[M–H]⁺	287	Cyanidin-3-O-glucoside	Anthocyanin	✓	✓	✓
32.1	561	[M–H]⁺	399	Vitisin A	Pyranoanthocyanin	✓	✓	✓
32.2	479	[M–H]⁺	317	Petunidin-3-O-glucoside	Anthocyanin	✓	✓	✓
33.0	479	[M–H]⁻	316	Myricetin-3-O-glucoside	Flavonol	✓	✓	✓
34.0	493	[M–H]⁺	331	Malvidin-3-O-glucoside	Anthocyanin	✓	✓	✓
34.1	463	[M–H]⁺	301	Peonidin-3-O-glucoside	Anthocyanin	✓	✓	✓
34.9	477	[M–H]⁻	301	Quercetin-3-O-glucuronide	Flavonol	✓	✓	✓
34.9	507	[M–H]⁺	303	Delphinidin-3-O-6″-*p*-acetylglucoside	Anthocyanin	✓	✓	✓
36.4	707	[M–H]⁺	399	10-carboxypyranomalvidin-3-6″-*p*-coumaroylglucoside	Pyranoanthocyanin	✓	✓	✓
36.9	521	[M–H]⁺	317	Petunidin-3-O-6″-*p*-acetylglucoside	Anthocyanin	✗	✓	✓
36.9	507	[M–H]⁻	345	Syringetin-3-O-glucoside	Flavonol	✓	✓	✓
37.1	491	[M–H]⁺	287	Cyanidin-3-O-6″-*p*-acetylglucoside	Anthocyanin	✗	✗	✓
39.6	317	[M–H]⁻	179, 151, 137	Myricetin	Flavonol	✓	✓	✓
39.2	535	[M–H]⁺	331	Malvidin-3-O-6″-*p*-acetylglucoside	Anthocyanin	✗	✓	✓
39.5	505	[M–H]⁺	301	Peonidin-3-O-6″-*p*-acetylglucoside	Anthocyanin	✗	✓	✓
39.8	611	[M–H]⁺	303	Delphinidin-3-O-6″-*p*-coumaroylglucoside	Anthocyanin	✓	✓	✓

Table 2. *Cont.*

Retention Time (min)	*m/z*	Ionic Species	Fragment Ions	Putative Identification	Phenolic Subclass	Grape Marc	Red Wine Lees	Port Wine Lees
41.9	595	[M–H]$^+$	287	Cyanidin-3-O-6″-*p*-coumaroylglucoside	Anthocyanin	✓	✓	✓
41.9	625	[M–H]$^+$	317	Petunidin-3-O-6″-*p*-coumaroylglucoside	Anthocyanin	✓	✓	✓
43.9	301	[M–H]$^-$	179,151, 121, 107	Quercetin	Flavonol	✓	✓	✓
43.8	639	[M–H]$^+$	331	Malvidin-3-O-6″-*p*-coumaroylglucoside	Anthocyanin	✓	✓	✓
44.3	609	[M–H]$^+$	301	Peonidin-3-O-6″-*p*-coumaroylglucoside	Anthocyanin	✓	✓	✓
48.2	285	[M–H]$^-$	125	Kaempferol	Flavonol	✓	✓	✓
49.2	315	[M–H]$^-$	300, 247, 215, 165, 141	Rhamnetin	Flavonol	✗	✓	✓

3.1.3. Antioxidant Activity Characterization

Phenolics are well known for their antioxidant activity, which is an important feature that determines the relevance of these compounds for cosmetic applications, given that ROS are the driving causes of skin ageing. Therefore, the extracts under study were submitted to three complementary antioxidant assays, aiming at assessing the antioxidant capacities of the samples towards different biologically relevant radical species, namely hydroxyl ($^\bullet$OH) and peroxyl (ROO$^\bullet$) radicals. Hydroxyl radicals are primarily generated from Fenton-like reactions, in which metal ions are oxidized by hydrogen peroxide (Metal^{2+} + H$_2$O$_2$→Metal^{3+} + $^\bullet$OH + OH$^-$), and are highly reactive species that are able to oxidize numerous biomolecules, including membrane lipids by initiating lipid peroxidation through the abstraction of hydrogen atoms from unsaturated fatty acids, resulting in the generation of peroxyl radicals (ROO$^\bullet$) [11].

The antioxidant activity results summed up in Table 1 show that in all extracts, there are compounds exerting the three types of antioxidant activity: Peroxyl and hydroxyl radicals scavenging, and transition metal ion chelation. The red wine lees extracts (RW and MW RW) presented significantly better results than all the other extracts in all three assays, with antioxidant capacity >3000 µmol TE/g extract in ORAC and HOSC, and HORAC values >1900 µmol CAE/g extract (Table 1). In the case of the Port wine lees extracts, MW P had significantly higher potential for peroxyl and hydroxyl radical scavenging as well as metal ion chelation, presenting higher ORAC, HOSC and HORAC values than P. Among the grape marc extracts, MW80 GM revealed lower antioxidant capacity than GM in HOSC and HORAC but not in ORAC.

The ORAC, HOSC and HORAC values correlated well with TPC, TAC and the relative amounts of flavonols for all samples, with R^2 values ≥0.95. Although the higher amounts of phenolics correspond to a more promising antioxidant activity, the different proportions of phenolic subclasses in the extracts may also lead to different activities towards distinct oxidant sources. For instance, flavonoids are generally more capable of inactivating peroxyl radicals than small phenolic antioxidants, whereas monohydroxybenzoic acids are very effective in the inactivation of hydroxyl radicals [42]. However, in this work, the antioxidant activity results seem to correlate well ($R^2 > 0.8$) with most of the compounds identified in the studied samples.

Similar to what has been observed for TPC, the ORAC values for the grape marc extracts were much lower than the ones described by Álvarez [17] (a decrease higher than 60% was observed for the two extracts), which may be justified, once again, by different vintages (2014 versus 2015). Nonetheless, the MW pretreated grape marc yielded the same ORAC values as the respective conventional extract, which is in agreement with Álvarez [17]. Regarding Port and red wine lees, the conventional extracts presented the same ORAC values as those reported in previous work [15]. Although for red wine lees, the extract obtained following MW pretreatment did not reveal a significantly higher ORAC value than the corresponding conventional extract, in the case of Port wine lees, the MW-pretreatment increased ORAC. Regarding HOSC and HORAC, similar results as the ones obtained for red wine lees were found in the literature for a similar matrix (ageing wine lees) [40].

The antioxidant activity values (as well as TPC and TAC) found in the literature for grape marc and wine lees of different origin as the ones used in this work are highly variable, and a comparison with the ones obtained herein is not easy because these determinations greatly depend on the grape variety, maturation stage, environmental conditions during grape growth, vinification parameters, and extraction procedure.

3.2. Screening of the Cosmetic Potential of Wine Lees and Grape Marc Extracts

3.2.1. Anti-Hyperpigmentation Activity

Inhibition of Tyrosinase

Along with ageing, pigmentation disorders tend to appear in the skin, which has attracted the attention of cosmetic industries and led the quest to find compounds with anti-hyperpigmentation potential. These pigmentation lesions are caused by alterations resulting in the accumulation of melanin, thus the inhibition of melanin production is the most explored approach in this field. As tyrosinase is the rate-limiting enzyme in melanin synthesis, it is a promising target for the development of skin-whitening cosmetic products. Due to their aromatic structural features, phenolics bear some similarities to tyrosine, the substrate of tyrosinase that initiates the synthesis of melanin. Hence, phenolics are potential analogs of tyrosine that can act as competitive inhibitors. In addition, tyrosinase contains a copper ion in its active site, and certain phenolics have the ability to chelate transition metal ions [43].

All tested extracts showed a dose-dependent inhibiting effect on tyrosinase, which allowed for the determination of the IC_{50} values (Table 3). The wine lees extracts presented lower IC_{50} values ($\leq$1.06 mg extract/mL) than the grape marc extracts ($\geq$4 mg extract/mL), which means that the former are more potent inhibitors of tyrosinase than the latter. Amongst the wine lees matrices, red wine lees (RW and MW RW) showed the best results, with the highest potential for tyrosinase inhibition ($IC_{50} \leq$ 0.2 mg extract/mL). The capacity of the extracts to inhibit tyrosinase does not correlate directly with TPC, TAC or the relative amount of flavonols ($R^2 \leq$ 0.64), suggesting that the capacity to inhibit tyrosinase relies more on the presence of specific compounds rather than the overall amount of phenolics. For instance, GM exhibited a TPC almost two-fold higher than MW80 GM (83.9 versus 45.9 mg GAE/g extract), yet these two extracts revealed the same IC_{50} towards tyrosinase.

Table 3. The IC_{50} values of extracts towards tyrosinase, elastase and MMP-1. The results identified with different letters (a to d) in the same column are statistically different (*p*-value $\leq$ 0.05).

		Anti-hyperpigmentation Activity	Anti-ageing Activity	
	Extract	**IC_{50} Tyrosinase (mg Extract/mL)**	**IC_{50} Elastase (mg Extract/mL)**	**IC_{50} MMP-1 (mg Extract/mL)**
Wine lees	RW	0.20 ± 0.01 [a]	0.17 ± 0.01 [a]	0.22 ± 0.01 [a]
	MW RW	0.14 ± 0.01 [a]	0.11 ± 0.00 [a]	0.21 ± 0.01 [a]
	P	1.06 ± 0.07 [b]	1.92 ± 0.09 [b]	1.25 ± 0.03 [b]
	MW P	0.62 ± 0.04 [a,b]	0.83 ± 0.04 [c]	0.65 ± 0.03 [a,c]
Grape marc	GM	4.03 ± 0.14 [c]	0.87 ± 0.03 [c]	1.08 ± 0.08 [b,c]
	MW80 GM	4.00 ± 0.14 [c]	3.43 ± 0.11 [d]	1.16 ± 0.06 [b]

RW—red wine lees conventional extract; MW RW—MW-pretreated red wine lees extract; P—Port wine lees conventional extract; MW P—MW-pretreated Port wine lees extract; GM—grape marc conventional extract; MW80 GM—MW-pretreated grape marc extract (max. temp. 80 °C).

The inhibition of tyrosinase may take place by the direct interference of phenolics with the active site, the allosteric interactions leading to conformational changes or the loss of function, or by copper ion chelation. Flavonoids possess metal-binding motifs in their structure [43], hence these phenolics may play a significant role in tyrosinase inhibition. The areas of the peaks obtained from the mass spectra, corresponding to the compounds identified in the samples, allowed for the calculation of correlation coefficients between the amount of each compound and the effectiveness of the extract in inhibiting tyrosinase. It was found that, although the correlations were not very high with any of the identified compounds, tyrosinase inhibition correlated better with syringetin-3-O-glucoside (R^2 = 0.69), myricetin (R^2 = 0.66) and malvidin-3-O-glucoside (R^2 = 0.65) than with all the other identified compounds ($R^2 \leq$ 0.48). In the literature, myricetin has already been described as a tyrosinase inhibitor [44]. Nevertheless, other compounds present in the extracts, in particular other flavonoid aglycones, may be contributing as well for the exhibited tyrosinase inhibitory activity, since quercetin, kaempferol, and catechins, among other, are also reported to be effective tyrosinase inhibitors [44].

Kojic acid is a well-studied inhibitor of tyrosinase that is widely used as a reference for a comparison with novel potential inhibitors of the enzyme [45]. Kojic acid's IC_{50} was 0.03 mg/mL, and the red wine lees extracts effectiveness came remarkably close, particularly MW RW, presenting an IC_{50}

value only of approximately five-fold higher than kojic acid, which is more promising than several other plant extracts [46].

3.2.2. Anti-ageing Activity

Inhibition of Elastase

Elastases contribute to the reduced amount of elastin in the skin by degrading not only the elastic fibers therein, but also newly formed elastin, hampering its correct assembly into functional elastic fibers. Moreover, elastases are broadly specific enzymes being also able to degrade other ECM proteins. Therefore, studying the capacity of the extracts to inhibit elastase is relevant for validating their anti-ageing potential. The IC_{50} values are displayed in Table 3 and, following the trend of previous assays, the red wine lees extracts presented the lowest IC_{50} values ($\leq$0.17 mg extract/mL). The correlations of elastase IC_{50} values with TAC and the relative amounts of flavonols were not very high ($R^2 \leq 0.53$) and, although the relationship between IC_{50} and TPC is not linear ($R^2 = 0.66$), there is a direct correspondence between these two parameters, which suggests that the amount of phenolics determines the inhibitory capacity of the extracts. In fact, it has been reported that higher TPC leads to stronger inhibition of elastase [10]. The inhibition of elastase relies on van der Waals (vdW) interactions and hydrogen bonds between the enzyme and the inhibitor, thus phenolics with a larger number of potential interaction sites, including aromatic rings for vdW interactions and hydroxyl groups for hydrogen bonding, are more likely to better inhibit elastase. The structural features playing an important part in the inhibitory capacity of phenolics towards elastase are the galloyl moiety, the degree of polymerization in the case of procyanidins, and hydroxylation of the structure [47,48]. However, the presence of glycosidic groups in the phenolic structure was found to preclude the inhibitory interaction with the enzyme, due to steric hindrance, suggesting that flavonoid aglycones may be more relevant for the elastase inhibitory capacity of the extracts than their respective glycoside derivatives [10,48]. Wittenauer [10] reported that gallic acid, catechin and procyanidins exhibited inhibitory activity against elastase, and Sartor [48] studied the effect of several compounds for this purpose, including myricetin and quercetin. Therefore, the authors suggest that gallic acid, catechin, epicatechin, procyanidin dimers and trimer, as well as the flavonol aglycones identified in the extracts (myricetin, quercetin, kaempferol and rhamnetin) are important in the capacity of winemaking waste streams extracts to inhibit elastase activity. Nonetheless, the IC_{50} values did not correlate particularly well with any specific compound ($R^2 \leq 0.53$), reinforcing the fact that the overall TPC is the most relevant feature for elastase inhibition.

Inhibition of MMP-1

Matrix metalloproteinase-1 (MMP-1), also known as interstitial collagenase, is one of the most important enzymes participating in the process of skin ageing. MMP-1 can initiate the degradation of fibrillar types of collagen, such as collagens I and III which are the predominant forms existing in skin, paving the way for further degradation by other enzymes. MMP-1 can also cleave non-fibrillar collagen and other ECM constituents. MMPs have a conserved methionine and a zinc-binding motif in their active site, as well as a similar fold [49], thus the inhibitory capacity of the extracts against MMP-1 may provide an insight into their capacity to inhibit other MMPs. The inhibition of collagenase can be achieved by either unspecific non-covalent interactions with amino acid side chains leading to conformational changes, interaction with the binding site, or complexation of the zinc ion (Zn^{2+}) present in the catalytic site [10]. The most effective inhibitor would combine both metal ion chelation and interactions with the protein through vdW forces and hydrogen bonds.

The IC_{50} values of MMP-1 inhibition by the extracts are displayed in Table 3. Once again, red wine lees presented the best results in terms of inhibitory capacity towards MMP-1 ($IC_{50} \leq 0.22$ mg extract/mL). A considerably linear relationship ($R^2 = 0.93$) between TPC and the IC_{50} values for MMP-1 is observed, which makes sense considering the lack of specificity involved in collagenase inhibition. Additionally, the correlations with TAC and with the relative amounts of flavonols were also good ($R^2 = 0.92$ and 0.89, respectively). In fact, several compounds from different phenolic subclasses have been shown to inhibit collagenase and/or other MMPs. Among them, catechins and procyanidins, gallic acid, delphinidin aglycone, myricetin, quercetin, and kaempferol. Further, the important structural features contributing to the inhibition of MMPs include the presence of galloyl moieties, polyhydroxylation of the flavonoid backbone, planarity of the molecule, and the presence of metal-binding motifs have been described [48]. Interestingly, the IC_{50} values for MMP-1 showed particularly good correlations with the compounds from different phenolic subclasses, namely caftaric acid ($R^2 = 0.89$) belonging to the class of phenolic acids, the anthocyanins malvidin- ($R^2 = 0.92$) and peonidin-3-O-glucoside ($R^2 = 0.89$), and the flavonols, myricetin ($R^2 = 0.88$) and quercetin ($R^2 = 0.84$). Therefore, these might be the main compounds responsible for the MMP-1 inhibitory capacity of the tested extracts, by being able to not only establish vdW interactions and hydrogen bonds with the protein, but also to chelate transition metal ions, including Zn^{2+}.

3.2.3. Cellular Antioxidant Activity

The chemical antioxidant assays are useful tools for an initial antioxidant activity screening of compounds or natural extracts. However, these methods have some limitations concerning the prediction of the antioxidant activity of tested samples in a biological environment. The parameters like bioavailability, cellular uptake, and metabolism are not taken into consideration in chemical assays given the simplicity of these systems. For instance, two compounds may have similar antioxidant activities as determined by chemical assays, yet one may be more promising than the other when applied in a biological context because of its availability at the site of action. The cellular antioxidant assays comprise of some of the complexity of biological systems, namely cellular uptake, subcellular location, and metabolism [28]. Since keratinocytes and fibroblasts are the predominant cell types encountered in the skin, representing the epidermal and dermal layers, respectively, the cellular assays were based on keratinocyte (HaCaT) and fibroblast (HFF, CCD-1112Sk) cell lines. These cell types are responsible for skin integrity and, when affected by senescence or oxidative stress, are leading players in the emergence of the aged skin phenotype.

Only the most promising extracts from each raw material were chosen to proceed to cell-based assays, and the selection was based on the results obtained in chemical and enzymatic assays. These results are summarized in Table 4, in which the extracts are rated according to their performance in each assay. MW P and MW RW were chosen due to the overall better results than the respective conventional extracts in the previous experiments. Concerning the grape marc extracts, GM was generally more promising than MW80 GM. However, MW80 GM showed a higher TAC than GM (Table 1), thus both GM and MW80 GM were selected for cell-based assays.

Primarily, the potential cytotoxicity of the extracts was evaluated in both HaCaT and HFF (Figures S1 and S2) in order to select the non-toxic concentrations for further studies. Then, as a first approach, the capacity of the extracts to inhibit endogenous ROS was evaluated after an incubation period of 1 h. Several concentrations of each extract were tested, and a dose-dependent relationship was observed for all the extracts in both cell lines and incubation periods (Figure S3). However, for comparison purposes, only the common concentration amongst all extracts (0.25 mg extract/mL) is shown in Figure 2. The differences between the same concentrations of different extracts may be justified by their TPC as well as their diverse composition, presenting distinct amounts of phenolics with the ability to permeate or interact with cell membranes. In fact, the correlations between the inhibition of endogenous ROS production and TPC, TAC and the relative amount of flavonols were fairly good ($R^2 \geq 0.91$ in HaCaT, and $R^2 \geq 0.71$ in HFF). Nevertheless, MW RW was the most promising extract, resulting in protection percentages of $\geq 32\%$, as opposed to $\leq 24\%$ observed for other extracts, in both cell lines.

Table 4. Summary of the results obtained by the extracts in chemical and enzymatic assays, in terms of phytochemical composition, antioxidant, anti-hyperpigmentation and anti-ageing activities. For each assay, the extracts were rated based on their percentage relative to the mean. Phytochemical composition and antioxidant activity: - for 0–50%; + for 50–100%; ++ for 100–150%; +++ for 150–200%; ++++ for 200–250%; +++++ for >250%. Enzymatic assays: +++++ for 0–10%; ++++ for 10–25%; +++ for 25–50%; ++ for 50–100%; + for 100–200%; - for >200%.

		Phytochemical Composition		Antioxidant Activity			Anti-hyperpigmentation Activity	Anti-ageing Activity	
	Extract	TPC	TAC	ORAC	HOSC	HORAC	Tyrosinase	Elastase	MMP-1
Wine lees	RW	+++	++++	++++	++++	++++	+++++	++++	+++
	MW RW	++++	+++++	+++++	+++++	+++++	+++++	+++++	++++
	P	+	+	-	-	+	+++	+	+
	MW P	+	+	+	+	++	+++	++	++
Grape marc	GM	+	-	-	-	-	+	++	+
	MW80 GM	-	-	-	-	-	+	-	+

RW—red wine lees conventional extract; MW RW—MW-pretreated red wine lees extract; P—Port wine lees conventional extract; MW P—MW-pretreated Port wine lees extract; GM—grape marc conventional extract; MW80 GM—MW-pretreated grape marc extract (max. temp. 80 °C).

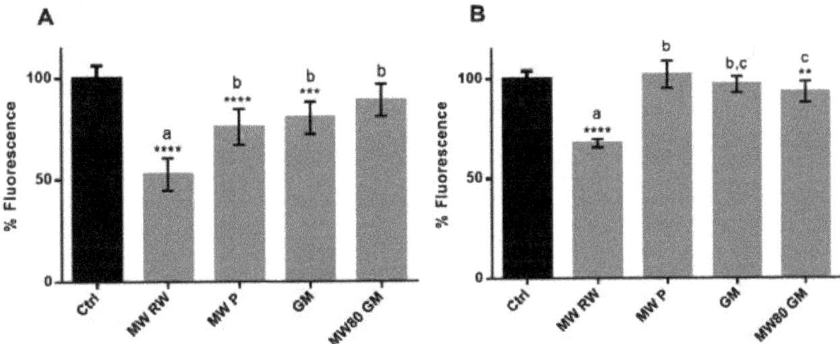

Figure 2. Pre-incubation of cells with 0.25 mg/mL of extracts for 1 h – effect on endogenous ROS. (**A**) HaCaT; (**B**) HFF. The symbol * indicates significance relative to the control (* p-value ≤ 0.05, ** p-value ≤ 0.01, *** p-value ≤ 0.001, **** p-value ≤ 0.0001). Statistical differences (p ≤ 0.05) between the samples are identified with different letters. MW RW—MW-pretreated red wine lees extract; MW P—MW-pretreated Port wine lees extract; GM—grape marc conventional extract; MW80 GM—MW-pretreated grape marc extract (max. temp. 80 °C).

In order to better understand the potential protective effects of the extracts towards keratinocytes and fibroblasts, further studies were performed using an oxidative stress inducer, TBHP, which is a more stable alkyl derivative of H_2O_2 that can initiate radical reactions, leading to damage of biomolecules. The effects of the extracts on TBHP-induced ROS was assessed in two different conditions: Pre-incubation of cells with the extracts prior to the addition of the stressor, and the co-incubation of the extracts with the stressor. Pre-incubation may reflect a preventive action, whereas co-incubation is more representative of a possible therapeutic approach. Several concentrations of the extracts were tested, and in both experiments a dose-dependent effect was observed (Figures S4 and S5). The results for the common concentration amongst the extracts (0.25 mg extract/mL) are presented in Figure 3. When comparing the two approaches, it is clear that the prevention of ROS formation is much more effective when the extracts are co-incubated with the stressor, with all the samples presenting a significant (p ≤ 0.0001) decrease of ROS in contrast with the untreated control (Figure 3C,D). These findings suggest that there must be compounds in the extracts that are not capable of permeating the cell membrane, and therefore their antioxidant properties can only be noticed when the induction of oxidative stress occurs in the presence of the extracts. In fact, certain phenolics may not be within the specific structural limitations required for membrane permeation, and therefore do not reach intracellular space. These observations are consistent with those reported elsewhere, in which *Opuntia ficus-indica* extracts [50] and traditional Portuguese cherry extracts [51] reveal a stronger antioxidant activity in co-incubation conditions rather than in the pre-incubation approach. The high molecular weight polyhydroxylated phenolics, such as certain flavonoids and their respective conjugates, might be among the compounds responsible for the distinction between the pre-incubation and co-incubation results, due to their difficulty permeating membranes [52]. Indeed, the inhibition of TBHP-induced ROS correlates better with TPC, TAC and the relative amount of flavonols in the case of co-incubation ($R^2 \geq 0.92$ in both cell lines) rather than in pre-incubation ($R^2 \leq 0.63$ in both cell lines). This makes sense, since in the co-incubation approach, all the compounds are present in the moment of stress induction, even those not able to permeate cells.

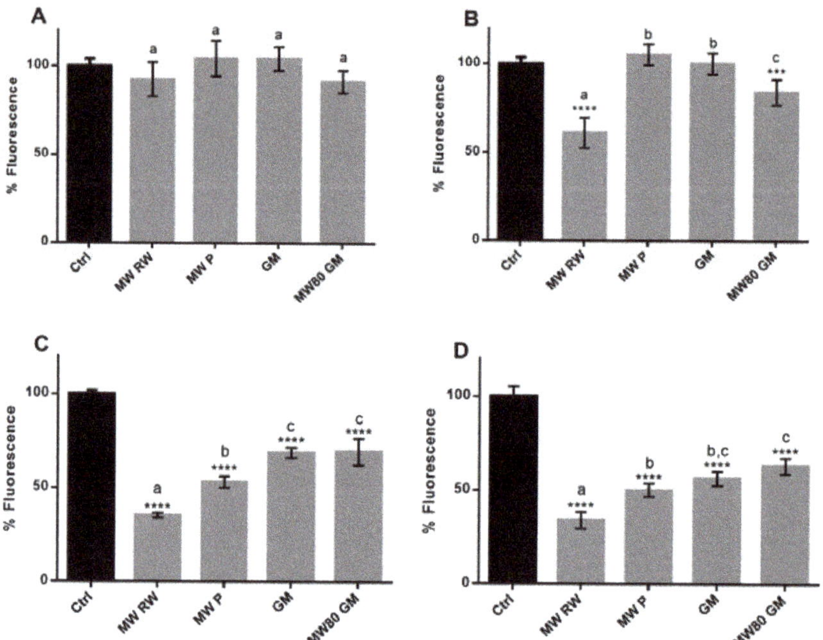

Figure 3. Pre-incubation of cells with 0.25 mg/mL of extracts for 1 h prior to addition of TBHP – effect on induced ROS – in HaCaT (**A**) and HFF (**B**); co-incubation of HaCaT (**C**) and HFF (**D**) with extracts and TBHP for 1 h. The symbol * indicates significance relative to the control (*** *p*-value $\leq$ 0.001, **** *p*-value $\leq$ 0.0001). Statistical differences ($p \leq 0.05$) between the samples are identified with different letters. MW RW—MW-pretreated red wine lees extract; MW P—MW-pretreated Port wine lees extract; GM—grape marc conventional extract; MW80 GM—MW-pretreated grape marc extract (max. temp. 80 °C).

The assessment of the protective effects of natural extracts against oxidative damage caused by an oxidative stress inducer is commonly performed in keratinocytes and fibroblasts [53,54]. In this approach, the cells were pre-incubated with the extracts for 24 h, and then a cytotoxic level of oxidative stress was induced with TBHP for 1 h. Finally, the MTS assay was performed, and the cell viability of the treated cells was compared to an untreated control where stress was also induced (Ctrl stress). In Figure 4, particularly in the case of HaCaT (Figure 4A), a biphasic dose-response can be seen, in which the extracts lead to beneficial effects until a specific concentration is reached, and then cytotoxicity emerges. This specific response is referred to as hormesis, and results from an adaptive response of an organism upon disruption in homeostasis caused by low doses of an exogenous factor, whereas at high doses the toxic effect prevails [55]. Although phenolics are renowned antioxidants, they can also have pro-oxidant effects in certain conditions, often presenting hormetic responses. For instance, although phenolics can chelate transition metal ions, in some cases a redox reaction takes place instead, yielding reactive metal ions even more prone to participate in the Fenton chemistry, as well as phenolic intermediates (phenoxyl radicals) with pro-oxidant properties [56]. However, in HFF (Figure 4B), this effect is not as evident as in HaCaT. In fact, all concentrations of the grape marc extracts seem to have potentiated the cytotoxic effect caused by TBHP, possibly because the amount of TBHP-induced ROS led to an exhaustion of the antioxidant capacity of the phenolics present in the extracts, triggering pro-oxidant effects that might have caused a further decrease in cell viability. Nonetheless, at a concentration of 0.25 mg extract/mL in both cell lines, the wine lees extracts either prevented TBHP-induced cytotoxicity, or did not present any differences relative to the control, yielding overall better results than the grape marc extracts.

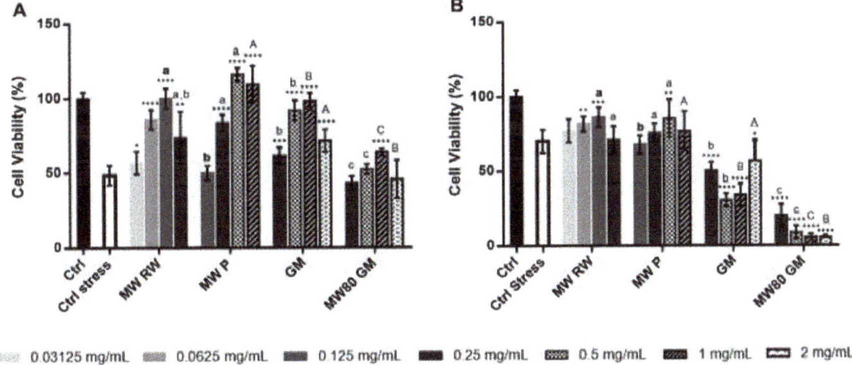

Figure 4. Pre-incubation of (**A**) HaCaT and (**B**) HFF with the extracts for 24 h – influence on cell viability upon TBHP-induced stress. The symbol * indicates significance relative to the control (* p-value ≤ 0.05, ** p-value ≤ 0.01, *** p-value ≤ 0.001, **** p-value ≤ 0.0001). The same concentrations of different extracts were compared (bold lowercase letters for 0.125 mg/mL; regular lowercase letters for 0.25 mg/mL; underlined lowercase letters for 0.5 mg/mL; regular uppercase letters for 1 mg/mL; underlined uppercase letters for 2 mg/mL). Statistically different results (p-value ≤ 0.05) are identified with different letters. MW RW—MW-pretreated red wine lees extract; MW P—MW-pretreated Port wine lees extract; GM—grape marc conventional extract; MW80 GM—MW-pretreated grape marc extract (max. temp. 80 °C).

4. Conclusions

As concluding remarks, the red wine lees extracts presented the highest phenolic and anthocyanin contents, leading to distinguishably better results than all the other tested extracts in terms of antioxidant activity, as measured by ORAC, HOSC and HORAC, tyrosinase, elastase and MMP-1 inhibitory capacity, and the protection of human skin cells (keratinocytes and fibroblasts) against oxidative stress. Moreover, the MW-pretreatment of raw materials seems to contribute to phenolic and anthocyanin richness of the extracts. Nonetheless, winemaking waste streams, in particular wine lees, are indeed valuable sources of natural bioactives with the potential for application in cosmeceutical products with antioxidant, skin whitening and anti-ageing effects.

Supplementary Materials: The following are available online at http://www.mdpi.com/2076-3921/8/9/355/s1, Figure S1: Cytotoxicity screening of the chosen extracts (24 and 48 h of incubation) in HaCaT, Figure S2: Cytotoxicity screening of the chosen extracts (24 and 48 h of incubation) in HFF, Figure S3: Pre-incubation of the cells with four concentrations of each extract for 1 h—effect on endogenous ROS, Figure S4: Pre-incubation of the cells with four concentrations of each extract for 1 h—effect on TBHP-induced ROS, Figure S5: Co-incubation of the cells with TBHP and four concentrations of each extract for 1 h. (A) HaCaT; (B) HFF.

Author Contributions: Conceptualization, S.R.-R., R.B.M., M.J.C. and A.A.M.; data curation, M.S.M.; formal analysis, M.S.M.; funding acquisition, R.R.-D., A.A., S.R.-R. and A.A.M.; investigation, M.S.M., R.R.-D., A.A., S.R.-R., and R.B.M.; methodology, R.R.-D., A.A. and R.B.M.; project administration, M.J.C; resources, M.R.B.; supervision, M.R.B., S.R.-R., R.B.M. and A.A.M.; validation, R.R.-D. and A.A.; writing—original draft, M.S.M.; writing—review & editing, M.S.M., M.R.B., S.R.-R. and A.A.M.

Funding: This work was supported by the European project WineSense (FP7-386 MC-IAPP), and by Junta de Castilla y León regional project VA040U16. Ana Álvarez would like to thank the Spanish Ministry of Education for her fellowship (FPU13/04678). Rut Romero acknowledges Junta de Castilla y León for her doctoral scholarship and Soraya Rodríguez acknowledges Junta de Castilla y León for her postdoctoral contract under VA040U16. A. A. Matias acknowledges the financial support received from the Portuguese Fundação para a Ciência e Tecnologia (FCT) through the PEst-OE/EQB/LA0004/2011 grant and by iNOVA4Health - UID/Multi/04462/2013. A. Matias also thanks FCT for her IF Starting Grant – GRAPHYT (IF/00723/2014).

Conflicts of Interest: The authors declare no conflicts of interest.

References

1. Bettini, O.; Sloop, C. Eu-27, Wine Annual: Wine Annual Report and Statistics 2015. *USDA Foreign Agricultural Service—Global Agricultural Information Network (GAIN)*, 2015; 1–28, GAIN Report Number: IT1512.

2. Ruggieri, L.; Cadena, E.; Martínez-Blanco, J.; Gasol, C.M.; Rieradevall, J.; Gabarrell, X.; Gea, T.; Sort, X.; Sánchez, A. Recovery of organic wastes in the Spanish wine industry. Technical, economic and environmental analyses of the composting process. *J. Clean. Prod.* **2009**, *17*, 830–838. [CrossRef]

3. Pinelo, M.; Arnous, A.; Meyer, A.S. Upgrading of grape skins: Significance of plant cell-wall structural components and extraction techniques for phenol release. *Trends Food Sci. Technol.* **2006**, *17*, 579–590. [CrossRef]

4. Bustamante, M.A.; Moral, R.; Paredes, C.; Pérez-Espinosa, A.; Moreno-Caselles, J.; Pérez-Murcia, M. Agrochemical characterisation of the solid by-products and residues from the winery and distillery industry. *Waste Manag.* **2008**, *28*, 372–380. [CrossRef] [PubMed]

5. Hashizume, H. Skin aging and dry skin. *J. Dermatol.* **2004**, *31*, 603–609. [CrossRef] [PubMed]

6. Gilchrest, B.A.; Blog, F.B.; Szabo, G. Effects of Aging and Chronic Sun Exposure on Melanocytes in Human Skin. *J. Investig. Dermatol.* **1979**, *73*, 141–143. [CrossRef] [PubMed]

7. Thakur, R.; Batheja, P.; Kaushik, D.; Michniak, B. Structural and Biochemical Changes in Aging Skin and Their Impact on Skin Permeability Barrier. In *Skin Aging Handbook: An Integrated Approach to Biochemistry and Product Development*; Dayan, N., Ed.; William Andrew: Norwich, NY, USA, 2008; pp. 55–90.

8. Sárdy, M. Role of Matrix Metalloproteinases in Skin Ageing. *Connect. Tissue Res.* **2009**, *50*, 132–138. [CrossRef] [PubMed]

9. Thring, T.S.; Hili, P.; Naughton, D.P. Anti-collagenase, anti-elastase and anti-oxidant activities of extracts from 21 plants. *BMC Complement. Altern. Med.* **2009**, *9*, 27–37. [CrossRef] [PubMed]

10. Wittenauer, J.; Mäckle, S.; Sußmann, D.; Schweiggert-Weisz, U.; Carle, R. Inhibitory effects of polyphenols from grape pomace extract on collagenase and elastase activity. *Fitoterapia* **2015**, *101*, 179–187. [CrossRef] [PubMed]

11. Birben, E.; Sahiner, U.M.; Sackesen, C.; Erzurum, S.; Kalayci, O. Oxidative Stress and Antioxidant Defense. *World Allergy Organ. J.* **2012**, *5*, 9–19. [CrossRef]

12. Yaar, M.; Gilchrest, B. Photoageing: Mechanism, prevention and therapy. *Br. J. Dermatol.* **2007**, *157*, 874–887. [CrossRef]

13. Fisher, G.J.; Kang, S.; Varani, J.; Bata-Csörgő, Z.; Wan, Y.; Datta, S.; Voorhees, J.J. Mechanisms of Photoaging and Chronological Skin Aging. *Arch. Dermatol.* **2002**, *138*, 1462–1470. [CrossRef] [PubMed]

14. Lin, Y.S.; Chen, H.J.; Huang, J.P.; Lee, P.C.; Tsai, C.R.; Hsu, T.F.; Huang, W.Y. Kinetics of Tyrosinase Inhibitory Activity Using Vitis vinifera Leaf Extracts. *BioMed Res. Int.* **2017**, *2017*, 5232680-5. [CrossRef] [PubMed]

15. Romero-Díez, R.; Matos, M.; Rodrigues, L.; Bronze, M.R.; Rodríguez-Rojo, S.; Cocero, M.; Matias, A. Microwave and ultrasound pre-treatments to enhance anthocyanins extraction from different wine lees. *Food Chem.* **2019**, *272*, 258–266. [CrossRef] [PubMed]

16. Perestrelo, R.; Silva, C.; Pereira, J.; Câmara, J.S. Wines: Madeira, Port and Sherry Fortified Wines—The Sui Generis and Notable Peculiarities. Major Differences and Chemical Patterns. In *Encyclopedia of Food and Health*; Caballero, B., Finglas, P.M., Toldrá, F., Eds.; Elsevier Ltd.: Amsterdam, The Netherlands, 2015; pp. 534–555.

17. Álvarez, A.; Poejo, J.; Matias, A.A.; Duarte, C.M.; Cocero, M.J.; Mato, R.B. Microwave pretreatment to improve extraction efficiency and polyphenol extract richness from grape pomace. Effect on antioxidant bioactivity. *Food Bioprod. Process.* **2017**, *106*, 162–170. [CrossRef]

18. Sólyom, K.; Solá, R.; Cocero, M.J.; Mato, R.B.; Alonso, M.J.C. Thermal degradation of grape marc polyphenols. *Food Chem.* **2014**, *159*, 361–366. [CrossRef] [PubMed]

19. Ainsworth, E.A.; Gillespie, K.M. Estimation of total phenolic content and other oxidation substrates in plant tissues using Folin–Ciocalteu reagent. *Nat. Protoc.* **2007**, *2*, 875–877. [CrossRef] [PubMed]

20. Singleton, V.L.; Orthofer, R.; Lamuela-Raventós, R.M. Analysis of total phenols and other oxidation substrates and antioxidants by means of folin-ciocalteu reagent. *Methods Enzymol.* **1998**, *299*, 152–178. [CrossRef]

21. Lee, J.; Durst, R.W.; Wrolstad, R.E. Determination of Total Monomeric Anthocyanin Pigment Content of Fruit Juices, Beverages, Natural Colorants, and Wines by the pH differential method: collaborative study. *J. AOAC Int.* **2005**, *88*, 1269–1278.

22. Huang, D.; Ou, B.; Hampsch-Woodill, M.; Flanagan, J.A.; Prior, R.L. High-Throughput Assay of Oxygen Radical Absorbance Capacity (ORAC) Using a Multichannel Liquid Handling System Coupled with a Microplate Fluorescence Reader in 96-Well Format. *J. Agric. Food Chem.* **2002**, *50*, 4437–4444. [CrossRef]

23. Moore, J.; Yin, J.J.; Yu, L. Novel Fluorometric Assay for Hydroxyl Radical Scavenging Capacity (HOSC) Estimation. *J. Agric. Food Chem.* **2006**, *54*, 617–626. [CrossRef]

24. Ou, B.; Hampsch-Woodill, M.; Flanagan, J.; Deemer, E.K.; Prior, R.L.; Huang, D. Novel Fluorometric Assay for Hydroxyl Radical Prevention Capacity Using Fluorescein as the Probe. *J. Agric. Food Chem.* **2002**, *50*, 2772–2777. [CrossRef] [PubMed]

25. Bolling, B.W.; Chen, Y.Y.; Kamil, A.G.; Chen, C.Y.O. Assay Dilution Factors Confound Measures of Total Antioxidant Capacity in Polyphenol-Rich Juices. *J. Food Sci.* **2012**, *77*, H69–H75. [CrossRef] [PubMed]

26. Chan, E.; Lim, Y.; Wong, L.; Lianto, F.; Wong, S.; Lim, K.; Joe, C.; Lim, T.; Chan, E.W.C. Antioxidant and tyrosinase inhibition properties of leaves and rhizomes of ginger species. *Food Chem.* **2008**, *109*, 477–483. [CrossRef]

27. Shimokawa, K.I.; Katayama, M.; Matsuda, Y.; Takahashi, H.; Hara, I.; Sato, H. Complexes of gelatinases and tissue inhibitor of metalloproteinases in human seminal plasma. *J. Androl.* **2003**, *24*, 73–77. [PubMed]

28. Wolfe, K.L.; Liu, R.H. Cellular Antioxidant Activity (CAA) Assay for Assessing Antioxidants, Foods, and Dietary Supplements. *J. Agric. Food Chem.* **2007**, *55*, 8896–8907. [CrossRef] [PubMed]

29. Rodríguez-Rojo, S.; Visentin, A.; Maestri, D.; Cocero, M. Assisted extraction of rosemary antioxidants with green solvents. *J. Food Eng.* **2012**, *109*, 98–103. [CrossRef]

30. Wang, L.; Weller, C.L. Recent advances in extraction of nutraceuticals from plants. *Trends Food Sci. Technol.* **2006**, *17*, 300–312. [CrossRef]

31. Fleuriet, A.; Macheix, J. Phenolic Acids in Fruits and Vegetables. In *Flavonoids in Health and Disease*, 2nd ed.; Rice-Evans, C.A., Packer, L., Eds.; Marcel Dekker Inc.: New York, NY, USA, 2003; pp. 1–41.

32. Peralbo-Molina, Á.; Luque deCastro, M.D. Potential of residues from the Mediterranean agriculture and agrifood industry. *Trends Food Sci. Technol.* **2013**, *32*, 16–24. [CrossRef]

33. Databank PhytoHub. Available online: www.phytohub.eu (accessed on 31 May 2018).

34. Databank Phenol-Explorer. Available online: www.phenol-explorer.eu (accessed on 31 May 2018).

35. Donato, P.; Rigano, F.; Cacciola, F.; Schure, M.; Farnetti, S.; Russo, M.; Dugo, P.; Mondello, L. Comprehensive two-dimensional liquid chromatography–tandem mass spectrometry for the simultaneous determination of wine polyphenols and target contaminants. *J. Chromatogr. A* **2016**, *1458*, 54–62. [CrossRef]

36. Blanco-Vega, D.; López-Bellido, F.J.; Alía-Robledo, J.M.; Hermosín-Gutiérrez, I. HPLC–DAD–ESI-MS/MS Characterization of Pyranoanthocyanins Pigments Formed in Model Wine. *J. Agric. Food Chem.* **2011**, *59*, 9523–9531. [CrossRef]

37. De La Torre, M.P.D.; De Castro, M.D.L.; Priego-Capote, F. Characterization and Comparison of Wine Lees by Liquid Chromatography–Mass Spectrometry in High-Resolution Mode. *J. Agric. Food Chem.* **2015**, *63*, 1116–1125. [CrossRef] [PubMed]

38. Schwarz, M.; Quast, P.; Von Baer, D.; Winterhalter, P. Vitisin A Content in Chilean Wines from Vitis viniferaCv. Cabernet Sauvignon and Contribution to the Color of Aged Red Wines. *J. Agric. Food Chem.* **2003**, *51*, 6261–6267. [CrossRef] [PubMed]

39. Wu, X.; Prior, R.L. Systematic Identification and Characterization of Anthocyanins by HPLC-ESI-MS/MS in Common Foods in the United States: Fruits and Berries. *J. Agric. Food Chem.* **2005**, *53*, 2589–2599. [CrossRef] [PubMed]

40. Romero-Díez, R.; Rodríguez-Rojo, S.; Cocero, M.J.; Duarte, C.; Matias, A.; Bronze, M. Phenolic characterization of aging wine lees: Correlation with antioxidant activities. *Food Chem.* **2018**, *259*, 188–195. [CrossRef] [PubMed]

41. Da Silva, M.A.; Salas, E.; Oliveira, J.; Teixeira, N.; De Freitas, V. Screening of Anthocyanins and Anthocyanin-Derived Pigments in Red Wine Grape Pomace Using LC-DAD/MS and MALDI-TOF Techniques. *J. Agric. Food Chem.* **2015**, *63*, 7636–7644.

42. Rice-Evans, C.A.; Miller, N.J.; Paganga, G. Structure-antioxidant activity relationships of flavonoids and phenolic acids. *Free Radic. Biol. Med.* **1996**, *20*, 933–956. [CrossRef]

43. Guo, M.; Perez, C.; Wei, Y.; Rapoza, E.; Su, G.; Bou-Abdallah, F.; Chasteen, N.D. Iron-binding properties of plant phenolics and cranberry's bio-effects†. *Dalton Trans.* **2007**, *43*, 4951–4961. [CrossRef]

44. Mukherjee, P.K.; Biswas, R.; Sharma, A.; Banerjee, S.; Biswas, S.; Katiyar, C. Validation of medicinal herbs for anti-tyrosinase potential. *J. Herb. Med.* **2018**, *14*, 1–16. [CrossRef]

45. Chang, T.S. An Updated Review of Tyrosinase Inhibitors. *Int. J. Mol. Sci.* **2009**, *10*, 2440–2475. [CrossRef]

46. Ya, W.; Chun-Meng, Z.; Tao, G.; Yi-Lin, Z.; Ping, Z. Preliminary screening of 44 plant extracts for anti-tyrosinase and antioxidant activities. *Pak. J. Pharm. Sci.* **2015**, *28*, 1737–1744.

47. Brás, N.F.; Gonçalves, R.; Mateus, N.; Fernandes, P.A.; Ramos, M.J.; De Freitas, V. Inhibition of Pancreatic Elastase by Polyphenolic Compounds. *J. Agric. Food Chem.* **2010**, *58*, 10668–10676. [CrossRef] [PubMed]

48. Sartor, L.; Pezzato, E.; Dell'Aica, I.; Caniato, R.; Biggin, S.; Garbisa, S. Inhibition of matrix-proteases by polyphenols: Chemical insights for anti-inflammatory and anti-invasion drug design. *Biochem. Pharmacol.* **2002**, *64*, 229–237. [CrossRef]

49. Nar, H.; Werle, K.; Bauer, M.M.; Dollinger, H.; Jung, B. Crystal structure of human macrophage elastase (MMP-12) in complex with a hydroxamic acid inhibitor. *J. Mol. Biol.* **2001**, *312*, 743–751. [CrossRef] [PubMed]

50. Matias, A.; Nunes, S.L.; Poejo, J.; Mecha, E.; Serra, A.T.; Madeira, P.J.A.; Bronze, M.R.; Duarte, C.M.M.; Bronze, M. Antioxidant and anti-inflammatory activity of a flavonoid-rich concentrate recovered from Opuntia ficus-indica juice. *Food Funct.* **2014**, *5*, 3269–3280. [CrossRef] [PubMed]

51. Serra, A.T.; Duarte, R.O.; Bronze, M.R.; Duarte, C.M. Identification of bioactive response in traditional cherries from Portugal. *Food Chem.* **2011**, *125*, 318–325. [CrossRef]

52. Tammela, P.; Laitinen, L.; Galkin, A.; Wennberg, T.; Heczko, R.; Vuorela, H.; Slotte, J.; Vuorela, P. Permeability characteristics and membrane affinity of flavonoids and alkyl gallates in Caco-2 cells and in phospholipid vesicles. *Arch. Biochem. Biophys.* **2004**, *425*, 193–199. [CrossRef] [PubMed]

53. Svobodová, A.; Rambousková, J.; Walterová, D.; Vostalová, J. Bilberry extract reduces UVA-induced oxidative stress in HaCaT keratinocytes: A pilot study. *BioFactors* **2008**, *33*, 249–266. [CrossRef] [PubMed]

54. Varasteh-kojourian, M.; Abrishamchi, P.; Matin, M.M.; Asili, J. Antioxidant, cytotoxic and DNA protective properties of *Achillea eriophora* DC. and *Achillea biebersteinii* Afan. extracts: A comparative study. *Avicenna J. Phytomed.* **2017**, *7*, 157–168. [CrossRef] [PubMed]

55. Calabrese, V.; Cornelius, C.; Salinaro, A.T.; Cambria, M.T.; Locascio, M.; Rienzo, L.; Condorelli, D.F.; Mancuso, C.; De Lorenzo, A.; Calabrese, E. The Hormetic Role of Dietary Antioxidants in Free Radical-Related Diseases. *Curr. Pharm. Des.* **2010**, *16*, 877–883. [CrossRef]

56. Bouayed, J.; Bohn, T. Exogenous Antioxidants—Double-Edged Swords in Cellular Redox State: Health Beneficial Effects at Physiologic Doses versus Deleterious Effects at High Doses. *Oxid. Med. Cell. Longev.* **2010**, *3*, 228–237. [CrossRef]

 *antioxidants* 

Article

Optimization of Sonotrode Ultrasonic-Assisted Extraction of Proanthocyanidins from Brewers' Spent Grains

Beatriz Martín-García [1], Federica Pasini [2], Vito Verardo [3,4,*], Elixabet Díaz-de-Cerio [3], Urszula Tylewicz [2], Ana María Gómez-Caravaca [1] and Maria Fiorenza Caboni [2,5]

[1] Department of Analytical Chemistry, Faculty of Sciences, University of Granada, Avd. Fuentenueva s/n, 18071 Granada, Spain
[2] Department of Agricultural and Food Sciences, University of Bologna, Piazza Goidanich 60, (FC) 47521 Cesena, Italy
[3] Department of Nutrition and Food Science, University of Granada, Campus of Cartuja, 18071 Granada, Spain
[4] Institute of Nutrition and Food Technology 'José Mataix', Biomedical Research Center, University of Granada, Avda del Conocimiento sn, 18100 Armilla, Granada, Spain
[5] Interdepartmental Centre for Agri-Food Industrial Research, Alma Mater Studiorum, Università di Bologna, via Quinto Bucci 336, 47521 Cesena (FC), Italy
* Correspondence: vitoverardo@ugr.es; Tel.: +34-958243863

Received: 28 June 2019; Accepted: 31 July 2019; Published: 6 August 2019

Abstract: Brewing spent grains (BSGs) are the main by-product from breweries and they are rich of proanthocyanidins, among other phenolic compounds. However, literature on these compounds in BSGs is scarce. Thus, this research focuses on the establishment of ultrasound-assisted extraction of proanthocyanidin compounds in brewing spent grains using a sonotrode. To set the sonotrode extraction up, response surface methodology (RSM) was used to study the effects of three factors, namely, solvent composition, time of extraction, and ultrasound power. Qualitative and quantitative analyses of proanthocyanidin compounds were performed using HPLC coupled to fluorometric and mass spectrometer detectors. The highest content of proanthocyanidins was obtained using 80/20 acetone/water (*v/v*), 55 min, and 400 W. The established method allows the extraction of 1.01 mg/g dry weight (d.w.) of pronthocyanidins from BSGs; this value is more than two times higher than conventional extraction.

Keywords: Box–Behnken design; proanthocyanidins; Brewers' spent grains; sonotrode ultrasonic-assisted extraction; HPLC-fluorometric detector (FLD)–MS

1. Introduction

Barley is the basic raw material for brewing. Phenolic compounds identified in barley include flavonoids, phenolic acids, and proanthocyanidins (PCs) [1,2]. There are more than 50 PCs in barley, comprising flavan-3-ol oligomers and their polymers [3]. The oligomers include dimers (prodelphinidin B3 and procyanidin B3), trimers, tetramers, and pentamers, while polymers are formed by oxidation and polymerization of simple flavan-3-ols [4]. Barley PCs ranged from 25 to 250 mg/100 g of grain [5–8]. Among them, proanthocyanidin trimers, such as catechin–gallocatechin–catechin (C–GC–C), prodelphinidin B3 and procyanidin B2 [9] are the most representative in barley. In addition, hops also contribute to the proanthocyanidin content in brewing spent grains (BSGs); in fact, according to several authors, this ingredient contains high amounts of catechin and procyanidins [10,11].

Furthermore, PCs showed anti-bacterial [12], anti-viral [13], anti-carcinogenic [14], anti-inflammatory [15], and cardioprotective effects [16]. Some studies demonstrated the potential of PCs for prevention or treatment of oxidative stress-associated diseases due to their antioxidant capacity [17]. In addition, PCs are easily extracted, affordable, and demonstrated low toxicity [17].

During the process of brewing, many BSGs are generated from barley grains after separation of the wort, and they consist of the residues from malted barley which could contain adjuncts (non-malt sources of fermentable sugars) such as wheat, rice, or maize and hop added during mashing [1]. Consequently, this by-product is rich in protein, fibers, arabinoxylans, and β-glucan, and also contains PCs in low concentration; thus, its reutilization could be useful for the food industry, and offers an opportunity for cereal-based baked and extruded products with acceptable sensory and nutritional characteristics [1].

In this sense, the challenge is to increase the efficient collection of PC-rich extracts with high bioactivity by the optimization of the extraction process. Thus far, conventional solid/liquid extraction was often used, employing as an extraction solvent a mixture of acetone and water in proportions from 50/50 to 80/20 [4,8,18,19] due to the large number of OH groups in PCs. In addition, bath-ultrasound-assisted extraction is the most used extraction technique. Some authors carried out pressurized solvent extraction, which is a static solid/liquid extraction with high pressure and eventually high temperature in stainless-steel extraction cells. Nevertheless, conventional extractions using ultrasonic-assisted extraction seem to be the best choice, since it is an economical technique, can be performed at atmospheric pressure and ambient temperature, and it could be developed on an ultrasound (US) bath or even with an US probe (or sonotrode) [20,21].

To carry out the determination of PCs in cereal, high-performance liquid chromatography (HPLC) is the analytical technique usually applied to this aim. In many instances, this technique was coupled to a diode array detector (DAD), fluorometric detector (FLD), and mass spectrometer detector (MSD) [8,22,23], or matrix-assisted laser desorption/ionization time-of-flight (MALDI-TOF) analysis [24].

In view of the above, the objective of this work was to evaluate the recovery of proanthocyanidins from BSGs by establishing a sonotrode ultrasonic-assisted extraction method. For that purpose, response surface methodology (RSM) was performed to evaluate extraction parameters with an experimental Box–Behnken design.

2. Materials and Methods

2.1. Samples

Brewers' spent grain (BSG) samples were obtained in a micro-brewing plant after pilsner beer production (Mastrobirraio, Cesena, Italy, 44°08′00″ north (N), 12°14′00″ east €).

2.2. Chemicals

HPLC-grade water and solvents were purchased from Merck KGaA (Darmstadt, Germany). Catechin was purchased from Sigma-Aldrich (St. Louis, MO).

2.3. Experimental Design

Response surface methodology (RSM) is the most popular tool for modeling. In RSM, statistical models and polynomial equations are always combined to provide an approximate relationship between the dependent and independent variables [25]. In the present work, a Box–Behnken design (BBD) with three factors was carried out in order to optimize the extraction parameters of proanthocyanidins in BSGs. The parameters of ultrasound-assisted extraction (US) can be divided into US parameters (ultrasound frequency, duration, acoustic power/intensity, and treatment mode) and non-US parameters (solvent type, solvent/sample ratio, particle size, temperature) [25]. In this work, the factors investigated were acetone/water (X1), time (X2), and potency (X3), with three levels for each factor, and the response variable (Y) was the sum of the total content of proanthocyanidins (PCs). The range for the percentage of acetone/water was chosen based on the conditions previously established in other works (50, 75, and

100%) [4,8]; the extraction time (5, 30, and 55 min) and the US power (80, 240, and 400 W) were the same as those previously used in a study where a sonotrode US was employed to optimize these parameters for the extraction of phenolic compounds from *Psidium guajava* L. leaves [26]. The design consisted of 15 combinations including three center points (Table 1), and the experiments were randomized to maximize the effects of unexplained variability in the observed response, due to extraneous factors.

The determination of optimal US sonotrode parameters was carried out using STATISTICA 7.0 (2002, StatSoft, Tulsa, OK).

2.4. Extraction of Proanthocyanidins from Brewers' Spent Grains by Sonotrode Ultrasonic Extraction

The extraction was achieved with a US sonotrode UP400St (Hielscher Ultrasonics GmbH, Teltow, Germany) and, during the extraction, an ice bath was used to avoid rises in temperature. The temperature ranged between 23 and 25 °C in all extractions, and it was measured with a thermometer at the end of each extraction. The percentage of acetone/water, the extraction time, and the US power were varied according to the experimental design. After the extraction, samples were centrifuged at 1000× g for 10 min; supernatants were collected, evaporated, and reconstituted in 1 mL of methanol/water (1/1, *v/v*). The final extracts were filtered through 0.2-μm polytetrafluoroethylene (PTFE) syringe filters and stored at −18 °C until the analyses.

2.5. Conventional Extraction of Proanthocyanidins

The results obtained by the US sonotrode at the optimal conditions were compared with a PC extract from BSGs obtained via conventional solid/liquid extraction. The extraction methodology was carried out according to Carciochi et al. [27]. Briefly, BSGs were subjected to mechanical agitation with a *w/v* ratio of 1/30, temperature of 80 °C, 72/28 ethanol/water (*v/v*), and an extraction time of 60 min.

2.6. Determination of Proanthocyanidins in Brewing Spent Grain Extracts by HPLC-FLD-MS Analysis

The separation of proanthocyanidins was performed on an Agilent 1200 Series HPLC system (Agilent Technologies, Santa Clara, CA) equipped with a binary pump delivery system, a degasser, an autosampler, and FLD and MS detectors (MSD, model G1946A, Santa Clara, CA, USA). A Develosil Diol 100 Å column (250 × 4.6 mm, 5 μm particle size) purchased from Phenomenex (Torrance, CA, USA) was used for the analyses.

All solvents were HPLC-grade and were filtered in a filter disc of 0.45 μm. According to Robbins et al. [28], the elution binary gradient consisted of CH3CN/HOAc, 98/2 (*v/v*) as solvent A, and $CH_3OH/H_2O/HOAc$ 95/3/2 *v/v/v* as solvent B. The analyses started with 7% of phase B from 0 to 3 min. Thus, solvent B was increased to 37.6% (from 3.1 to 57 min) and then to 100% B over the next 3 min for 7 min. After that, the initial condition was established, and they were maintained for 16 min. The injection volume was 5 μL and all the analyses were run at 35°C. Additionally, fluorescence detection was conducted with an excitation wavelength of 230 nm and an emission wavelength of 321 nm.

Moreover, identification of proanthocyanidins was carried out by HPLC-MS according to Verardo et al. [8]. Furthermore, quantification of PCs was done employing a calibration curve of (+)-catechin done from the limit of quantitation (LOQ) to 250 μg/mL (LOQ = 0.193 μg/mL). In addition, the quantification of dimers, trimers, tetramers, pentamers, and the polymers was done using the correction factors suggested by Robbins et al. [28].

3. Results and Discussion

3.1. Determination of Proanthocyanidin Compounds in Brewers' Spent Grains

Table 1 shows the sum of the total content of proanthocyanidins according to the experimental design (Table 1).

Table 1. Box-Behnken design (BBD) with the values of the sonotrode ultrasound (US) parameters with the experimental values for the dependent response of proanthocyanins (PCs) quantified by HPLC-fluorometric detector (FLD) in brewers' spent grain (BSG) extracts; d.w.—dry weight.

Experiment	Independent Factors			Dependent Factor
	X_1	X_2	X_3	Total ($\mu g \cdot g^{-1}$ d.w.)
1	50	5	240	540.04
2	100	5	240	548.25
3	50	55	240	690.90
4	100	55	240	802.25
5	50	30	80	547.91
6	100	30	80	849.32
7	50	30	400	601.43
8	100	30	400	792.07
9	75	5	80	796.40
10	75	55	80	977.69
11	75	5	400	993.15
12	75	55	400	1002.31
13	75	30	240	832.04
14	75	30	240	857.04
15	75	30	240	752.68

X1: acetone/water, X2: time, and X3: US power.

A total of 11 PCs were identified in BSGs according to their degree of polymerization and their mass spectra. As shown in Table 2 (and in Figure S1), the elution order depended on the number of flavan-3-ol units. Therefore, monomers eluted first and then the different oligomers eluted. In addition, for the same degree of polymerization, a higher degree of galloylation meant a higher retention time [8].

Table 2. Table of identification of proanthocyanidins from brewers' spent grain extracts by HPLC-MS; Rt—retention time.

Peak	Rt (min)	Compound	[M-H]⁻
1	6.7	Catechin/epicatechin	289
2	17.6	Procyanidin dimer	577
3	19.0	Prodelphinidin dimer	593
4	21.2	Prodelphinidin dimer II	593
5	24.4	Procyanidin trimer	865
6	26.8	Prodelphinidin trimer I (monogalloylated)	881
7	29.5	Prodelphinidin trimer II (digalloylated)	897
8	32.8	Procyanidin tetramer	1153
9	33.9	Prodelphinidin tetramer (digalloylated)	1457
10	36	Procyanidin pentamer	1441
11	51.7	Polymers (degree of polymerization >5)	

Moreover, quantification of PCs in brewing by-products was carried out using HPLC-FLD. The calibration curve of catechin was used to quantify the PCs. The correction factors were applied according to Robbins et al. [28]. The concentration values of PCs obtained in each experiment in the BBD are presented in Table 3. Briefly, the total content of PCs varied from 540.04 $\mu g \cdot g^{-1}$ dry weight (d.w.) to 1002.31 $\mu g \cdot g^{-1}$ d.w. Comparing the quantification of each compound, experiment 11, whose parameters of extraction were 75% acetone, 5 min, and 400 W of US power, recovered higher amounts of catechin/epicatechin, dimers, trimers, and tetramers than the rest of the experiments. Finally, the major concentrations of procyanidin pentamer, the polymer, and the total content of PCs were obtained in experiment 12 with 75% acetone, 55 min, and 400 W of US power.

Table 3. Table of quantification of proanthocyanidins from brewers' spent grain extracts by HPLC-FLD expressed as $\mu g \cdot g^{-1}$ d.w. UAE—ultrasound-assisted extraction; LOQ—limit of quantitation.

Proanthocyanidin Compounds	UAE 1	UAE 2	UAE 3	UAE 4	UAE 5	UAE 6	UAE 7	UAE 8	UAE 9	UAE 10	UAE 11	UAE 12	UAE 13	UAE 14	UAE 15
Catechin/epicatechin	8.34	9.17	10.16	9.71	8.05	10.03	10.07	10.37	9.59	10.33	10.41	8.41	9.62	9.53	8.89
Procyanidin dimer	50.08	70.49	52.50	85.90	40.06	73.45	44.02	82.34	57.47	76.36	100.92	64.17	98.56	88.97	73.94
Prodelphinidin dimer	22.68	33.01	26.09	25.96	30.44	38.86	31.93	43.95	49.16	57.03	38.74	25.68	31.04	33.97	31.60
Prodelphinidin dimer II	25.69	35.62	51.16	66.60	38.16	78.03	37.09	79.55	59.02	72.00	74.00	79.08	64.73	76.73	60.15
Procyanidin trimer	73.11	28.69	61.50	54.93	54.45	67.35	37.20	64.29	88.65	92.85	103.78	52.06	103.23	97.27	95.05
Prodelphinidin trimer I (monogalloylated)	35.58	73.86	56.78	97.85	49.08	101.98	45.60	95.27	92.53	122.39	121.94	81.68	98.98	107.81	83.69
Prodelphinidin trimer II (digalloylated)	<LOQ	48.58	<LOQ	82.52	<LOQ	80.67	<LOQ	71.26	79.53	92.77	83.62	75.12	65.34	78.03	59.54
Procyanidin tetramer	<LOQ	29.46	<LOQ	46.57	<LOQ	51.15	<LOQ	44.52	45.68	56.49	55.10	45.12	<LOQ	<LOQ	<LOQ
Prodelphinidin tetramer (digalloylated)	<LOQ	32.70	<LOQ	52.06	<LOQ	58.55	<LOQ	51.20	50.76	64.87	68.59	63.57	<LOQ	<LOQ	<LOQ
Procyanidin pentamer	<LOQ	17.64	<LOQ	26.50	<LOQ	28.01	<LOQ	19.34	24.84	35.28	30.44	42.78	<LOQ	<LOQ	<LOQ
Polymers	324.57	169.04	432.71	253.66	327.67	261.23	395.52	229.98	239.17	297.31	305.60	464.64	360.52	364.73	339.83
Total	540.04	548.25	690.90	802.25	547.91	849.32	601.43	792.07	796.40	977.69	993.15	1002.31	832.04	857.04	752.68

Proanthocyanidins were grouped as monomer, dimers, trimers, tetramers, pentamers, and polymers.

3.2. Fitting the Model

The regression model for the BBD was fitted employing the data from Table 1 in order to find the combined effect of extraction time, acetone/water ratio, and sonotrode US power on the response variable during the sonotrode US. For that, an analysis of variance (ANOVA) with 95% confidence level was employed to analyze the regression model and to evaluate the effect of the coefficients for each factor (linear and quadratic terms) and the interaction between them (cross-product term). In fact, the evaluation of the model was carried out according to the significance of the regression coefficients which are displayed in Table 4. According to other works, the level of significance could be fixed at $\alpha < 0.1$ in order to increase the number of significant terms [26]. In the present work, the model was analyzed at $\alpha < 0.05$ and $\alpha < 0.1$ The significant variables for the total content of PCs were the intercept (X_0) $(p = 0.000426)$, the linear effect of acetone/water (X_1) $(p = 0.058033)$ and its quadratic effect (X_{11}) $(p = 0.018319)$, the linear effect of time (X_2) $(p = 0.060966)$, and the quadratic effect of the power (X_{33}) $(p = 0.085914)$. Furthermore, ANOVA revealed that the model presented a high correlation between the factors and the response variables with a coefficient of determination (R^2) of 0.8999 (Table 4). In addition, the *p*-value of the regression model and the *p*-value of the lack-of-fit (LOF) were also used to verify the adequacy of the model. In fact, a high correlation term, a significant regression model $(p < 0.05)$, and a non-significant LOF $(p > 0.05)$ demonstrated the validity of the model (Table 4).

Table 4. Regression coefficients and ANOVA table.

Regression Coefficients	Total Proanthocyanidins
β_0	−1256.27 *
Linear	
β_1	53.07 **
β_2	−1.19 **
β_3	−0.68
Cross product	
β_{12}	0.04
β_{13}	−0.01
β_{23}	−0.01
Quadratic	
β_{11}	−0.33 *
β_{22}	0.06
β_{33}	0.00 **
R^2	0.8999
p (regression model)	0.0074
p (lack-of-fit)	0.3420

* Significant at $\alpha \leq 0.05$, ** significant at $\alpha \leq 0.1$; β_1: acetone/water ratio, β_2: time, β_3: US power, β_0: regression coefficient of mean.

3.2.1. Analysis of Response Surfaces

In order to determine the optimal value of each factor for the extraction of PCs from BSGs, response surfaces were plotted. Each pair of variables was depicted in three-dimensional surface plots, while the other factor was kept constant at a central level. Figure 1 shows the three-dimensional plots for the effects of acetone/water (% (*v/v*)) (X1) with time (X2), acetone/water (% (*v/v*)) (X1) with US power (X3), and time (X2) with US power (X3) on the concentration of the total content of PCs.

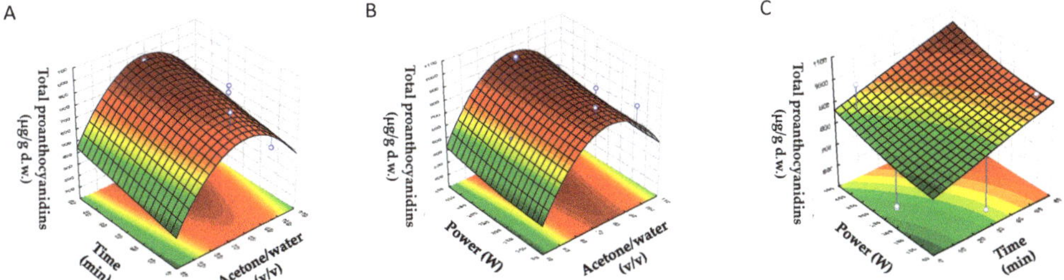

Figure 1. Response surface plots showing the combined effects of process variables for total proanthocyanidins: (**A**) acetone/water (% (*v/v*)) vs. time (min); (**B**) acetone/water (% (*v/v*)) vs. ultrasound (US) power (W); (**C**) time (min) vs. US power (W).

In Figure 1A,B, it can be observed that the response of the total content of PCs increased when the concentration of acetone increased at first. After that, a decrease in response was observed when the maximum response was achieved. This shape was a consequence of the quadratic effect of acetone, which had a negative value, showing that an increase in this parameter more than a certain value tended to decrease the response. For example, Figure 1A shows an increase in total concentration of PCs if the content of acetone rose until the maximum value (75–85%), for which the increase time caused a slight increase in the total concentration of PCs. Additionally, in Figure 1B, an increase in the content of total PCs up to 70–85% acetone was observed where it started to reduce, whereas the response increased slightly at 70–85% if the power increased. At last, Figure 1C shows the positive linear effect of time and power on the response; there was an increase in response with time and power.

3.2.2. Optimization of Sonotrode US Parameters

The optimal conditions were selected through the three-dimensional (3D) plots to obtain the highest content of PCs from BSGs, as shown in Table 5.

Table 5. Optimal conditions for sonotrode UAE.

Optimal Conditions	Sum of Proanthocyanidins (μg·g^{-1} d.w.)
Acetone/ water ratio (% (*v/v*))	80
Time (min)	55
US power	400
Predicted (μg·g^{-1} d.w.)	1012.7 ± 15.1
Obtained value (μg·g^{-1} d.w.)	1023.0 ± 8.9
Significant differences between predicted and obtained value	N.S.

N.S.: non-significant difference.

Briefly, optimal extraction conditions were 80% acetone/water (*v/v*), 55 min, and 400 W for US power. The final step of the RSM after selecting the optimal conditions was to verify the accuracy of the mathematical model. For that, an extraction at optimal conditions was done with the same methodology; the obtained value did not report significant differences with the predicted value.

According to the results, the maximum content of PCs was obtained at 80% acetone/water, because PCs with a high degree of polymerization were the most concentrated, and they were better extracted at a high percentage of acetone, since they were less polar than the other PCs, increasing their solubility in this solvent. Also, acetone was not an efficient solvent when used pure, showing good results when it was combined with water. This occurred due to increased solvation provided by the presence of water. Additionally, at a high time of extraction and maximum power, cell walls were disrupted, releasing proanthocyanidins from the cell constituents. The predicted values of the model were in accordance with the experimental data under the same conditions. In fact, no significant differences were noted between the two data.

3.3. Comparison between Conventional and Established Sonotrode Extraction

Table 6 displays the comparison between the extraction of flavan-3-ols using sonotrode US at the optimal conditions established by our model and that using conventional extraction carried out according to Carciochi et al. [27].

Table 6. Comparison of proanthocyanidin content using sonotrode and conventional extractions (µg/g d.w.).

Proanthocyanidin Compounds	Sonotrode Extraction	Conventional Extraction
Catechin/epicatechin	8.96 ± 0.23	3.89 ± 0.36
Procyanidin dimer	66.21 ± 1.10	21.34 ± 1.04
Prodelphinidin dimer	26.08 ± 0.29	10.25 ± 0.92
Prodelphinidin dimer II	80.43 ± 1.62	39.41 ± 1.37
Procyanidin trimer	53.19 ± 1.06	18.69 ± 2.06
Prodelphinidin trimer I (monogalloylated)	83.70 ± 2.12	42.16 ± 1.89
Prodelphinidin trimer II (digalloylated)	76.14 ± 0.98	35.47 ± 1.25
Procyanidin tetramer	47.09 ± 0.63	19.36 ± 0.47
Prodelphinidin tetramer (digalloylated)	65.22 ± 1.52	20.93 ± 1.12
Procyanidin pentamer	46.81 ± 1.70	18.71 ± 0.43
Polymers	469.21 ± 6.69	200.36 ± 2.89
Total	1023.04 ± 8.9	430.57 ± 3.62

According to the results obtained, the proposed methodology recovered 57.9% more total content of PCs than conventional extraction. Therefore, sonotrode ultrasound-assisted extraction is a more effective technique than conventional extraction for the recovery of PCs from BSGs. These data are in agreement with the data presented by Carciochi et al. [27].

Moreover, comparison with the literature is difficult because the information about the proanthocyanidin composition of BSGs is scarce. Comparing the values of proanthocyanidins obtained in this work with that obtained in barley samples, the contents of catechin, procyanidins, and prodelphynidins obtained in this work were on the same order of magnitude as those obtained in barley samples [4,8]. According to Moreira and co-workers [29], the present data also confirmed that light malt types as used for pilsner beer production contain high amounts of phenolic compounds.

In spite of proanthocyanidins being degraded at high temperatures during malting, where barley is milled, mixed with water in the mash tun, and the temperature of mash slowly increased from 37 to 78 °C to promote enzymatic hydrolysis of malt constituents [1], and during beer production, it was confirmed that a part of barley and hop proanthocyanidins still remain in the beer spent grains after beer production. Concentrations of catechin obtained at optimum sonotrode US conditions and in conventional extraction (8.96 ± 0.23 and 3.89 ± 0.36 mg·g^{-1} d.w., respectively) were higher than that reported by Ikram et al. [30] in brewers spent grain samples (1.08 ± 0.04 µg·100 g^{-1} d.w.). These differences could be because the catechin content of BSG varies according to barley variety, harvest time, malting and mashing conditions, and the quality and type of adjuncts added in the brewing process [1], but could also be due to the extraction method adopted for the proanthocyanidin extraction.

4. Conclusions

HPLC-FLD-MS was used for the determination of proanthocyanidins in brewers spent grains for the first time. A Box–Behnken experimental design was used in order to optimize the sonotrode ultrasound-assisted extraction parameters to obtain the maximum proanthocyanidin content from BSG. According to the model, the most important effect on the response came from the quadratic term of acetone/water ratio, followed by the linear term of acetone/water, the linear term of the time of extraction, and the quadratic term of US power. The highest value of proanthocyanidins was obtained at 80% acetone/water (*v/v*), 55 min, and 400 W. Finally, it was proven that sonotrode ultrasonic extraction is a more effective technique than conventional extraction method, providing a higher recovery of proanthocyanidins from BSG.

To conclude, BSGs represent a good raw material that could be used for the extraction of bioactive compounds or could be reused for the production of functional flours. In this way, further work will be done in order to validate this hypothesis.

Supplementary Materials: The following are available online at http://www.mdpi.com/2076-3921/8/8/282/s1, Figure S1: Separation of BSG proantocyanidins by HPLC-FLD.

Author Contributions: Conceptualization, V.V. and A.M.G.; data curation, B.M.G. and E.D.C.; investigation, B.M.G., F.P., and U.T.; supervision, V.V., A.M.G., and M.F.C.; writing—original draft, B.M.G.; writing—review and editing, F.P., V.V., E.D., U.T., A.M.G., and M.F.C.

Funding: This project was supported by the University of Granada (project PPJI2017.16).

Acknowledgments: Vito Verardo thanks the Spanish Ministry of Economy and Competitiveness (MINECO) for the "Ramon y Cajal" contract (RYC-2015-18795). Beatriz Martín García would like to thank the University of Granada for the "convocatoria de movilidad internacional de estudiantes de doctorado" grant.

Conflicts of Interest: The authors declare no conflicts of interest. The funders had no role in the design of the study; in the collection, analyses, or interpretation of data; in the writing of the manuscript, or in the decision to publish the results.

References

1. Gupta, M.; Abu-Ghannam, N.; Gallaghar, E. Barley for brewing: characteristic changes during malting, brewing and applications of its by-products. *Compr. Rev. Food Sci. Food Saf.* **2010**, *9*, 318–328. [CrossRef]

2. Hajji, T.; Mansouri, S.; Vecino-Bello, X.; Cruz-Freire, J.M.; Rezgui, S.; Ferchichi, A. Identification and characterization of phenolic compounds extracted from barley husks by LC-MS and antioxidant activity in vitro. *J. Cereal Sci.* **2018**, *81*, 83–90. [CrossRef]

3. Sharma, P.; Longvah, T. Chapter 2: Barley. In *Whole Grains: Processing, Product Development and Nutritional Aspects*; Mir, S.A., Manickavasagan, A., Shah, M.A., Eds.; CRC Press: Boca Raton, FL, USA, 2019; pp. 25–49.

4. Madigan, D.; McMurrough, I.; Smyth, M.R. Determination of proanthocyanidins and catechins in beer and barley by high-performance liquid chromatography with dual-electrode electrochemical detection. *The Anal.* **1994**, *119*, 863–868. [CrossRef]

5. Gómez-Caravaca, A.M.; Verardo, V.; Berardinelli, A.; Marconi, E.; Caboni, M.F. A chemometric approach to determine the phenolic compounds in different barley samples by two different stationary phases: A comparison between C18 and pentafluorophenyl core shell columns. *J. Chromatogr. A* **2014**, *1355*, 134–142. [CrossRef] [PubMed]

6. Gu, L.; Kelm, M.A.; Hammerstone, J.F.; Beecher, G.; Holden, J.; Haytowitz, D.; Gebhardt, S.; Prior, R.L. Concentrations of proanthocyanidins in common foods and estimations of normal consumption. *J. Nutr.* **2004**, *134*, 613–617. [CrossRef] [PubMed]

7. Holtekjolen, A.K.; Kinitz, C.; Knutsen, S.H. Flavanol and bound phenolic acid contents in different barley varieties. *J. Agric. Food Chem.* **2006**, *54*, 2253–2260. [CrossRef] [PubMed]

8. Verardo, V.; Cevoli, C.; Pasini, F.; Gómez-Caravaca, A.M.; Marconi, E.; Fabbri, A.; Caboni, M.F. Analysis of oligomer proanthocyanidins in different barley genotypes using high-performance liquid chromatography – fluorescence detection – mass spectrometry and near-infrared methodologies. *J. Agric. Food Chem.* **2015**, *63*, 4130–4137. [CrossRef] [PubMed]

9. Verardo, V.; Gómez-Caravaca, A.M.; Marconi, E.; Caboni, M.F. Air classification of barley flours to produce phenolic enriched ingredients: Comparative study among MEKC-UV, RP-HPLC-DAD-MS and spectrophotometric determinations. *LWT-Food Sci. Technol.* **2011**, *44*, 1555–1561. [CrossRef]

10. Li, H.J.; Deinzer, M.L. Proanthocyanidins in Hops. In *Beer in Health and Disease Prevention*; Preedy, V.R., Ed.; Accademic Press: Orlando, FL, USA, 2009; pp. 333–348.

11. Yamamoto, H.; Nagano, C.; Takeuchi, F.; Shibata, J.; Tagashira, M.; Ohtake, Y. Extraction of polyphenols in hop bract part discharged from beer breweries. *J. Chem. Eng. Jpn.* **2006**, *39*, 956–962. [CrossRef]

12. Mayer, R.; Stecher, G.; Wuerzner, R.; Silva, R.C.; Sultana, T.; Trojer, L.; Feuerstein, I.; Krieg, C.; Abel, G.; Popp, M.; et al. Proanthocyanidins: Target compounds as antibacterial agents. *J. Agric. Food Chem.* **2008**, *56*, 6959–6966. [CrossRef] [PubMed]

13. Xu, X.; Xie, H.; Wang, Y.; Wei, X. A-Type proanthocyanidins from lychee seeds and their antioxidant and antiviral activities. *J. Agric. Food Chem.* **2010**, *58*, 11667–11672. [CrossRef] [PubMed]

14. Nandakumar, V.; Singh, T.; Katiyar, S.K. Multi-targeted prevention and therapy of cancer by proanthocyanidins. *Cancer Lett.* **2008**, *269*, 378–387. [CrossRef] [PubMed]

15. Tatsuno, T.; Jinno, M.; Arima, Y.; Kawabata, T.; Hasegawa, T.; Yahagi, N.; Takano, F.; Ohta, T. Anti-inflammatory and anti-melanogenic proanthocyanidin oligomers from peanut skin. *Biol. Pharm. Bull.* **2012**, *35*, 909–916. [CrossRef] [PubMed]

16. Martín-Fernández, B.; de las Heras, N.; Valero-Muñoz, M.; Ballesteros, S.; Yao, Y.Z.; Stanton, P.G.; Fuller, P.J.; Lahera, V. Beneficial effects of proanthocyanidins in the cardiac alterations induced by aldosterone in rat heart through mineralocorticoid receptor blockade. *PLoS ONE* **2014**, *9*, 1–10. [CrossRef] [PubMed]

17. Yang, L.; Xian, D.; Xiong, X.; Lai, R.; Song, J.; Zhong, J. Proanthocyanidins against oxidative stress: From molecular mechanisms to clinical applications. *Biomed Res. Int.* **2018**, 8584136. [CrossRef] [PubMed]

18. Zuorro, A.; Iannone, A.; Lavecchia, R. Water–organic solvent extraction of phenolic antioxidants from brewers' spent grain. *Processes* **2019**, *7*, 126. [CrossRef]

19. Meneses, N.M.G.; Martins, S.; Teixeira, J.A.; Mussatto, S.I. Influence of extraction solvents on the recovery of antioxidant phenolic compounds from brewer's spent grains. *Sep. Purif. Technol.* **2013**, *108*, 152–158. [CrossRef]

20. Heng, M.Y.; Tan, S.W.; Yong, J.W.H.; Ong, E.S. Trends in analytical chemistry emerging green technologies for the chemical standardization of botanicals and herbal preparations. *Trends Analyt. Chem.* **2013**, *50*, 1–10. [CrossRef]

21. Galanakis, C.M. Recovery of high added value component from food wastes: Conventional, emerging technologies and commercialized applications. *Trend Food Sci. Tech.* **2012**, *26*, 68–87. [CrossRef]

22. Gu, L.; Kelm, M.A.; Hammerstone, J.F.; Beecher, G.; Holden, J.; Haytowitz, D.; Prior, R.L. Screening of foods containing proanthocyanidins and their structural characterization using LC-MS/MS and thiolytic degradation. *J. Agric. Food Chem.* **2003**, *51*, 7513–7521. [CrossRef]

23. Hellström, J.K.; Törrönen, A.R.; Mattila, P.H. proanthocyanidins in common food products of plant origin. *J. Agric. Food Chem.* **2009**, *57*, 7899–7906. [CrossRef] [PubMed]

24. Monagas, M.; Quintanilla-López, J.E.; Gómez-Cordovés, C.; Bartolomé, B.; Lebrón-Aguilar, R. MALDI-TOF MS analysis of plant proanthocyanidins. *J. Pharm. Biomed. Anal.* **2010**, *51*, 358–372. [CrossRef] [PubMed]

25. Tao, Y.; Sun, D.W. Critical reviews in food science and nutrition enhancement of food processes by ultrasound: A review. *Crit. Rev. Food Sci. Nutr.* **2015**, *55*, 570–594. [CrossRef] [PubMed]

26. Díaz-de-Cerio, E.; Tylewicz, U.; Verardo, V.; Fernández-Gutiérrez, A.; Segura-Carretero, A.; Romani, S. Design of sonotrode ultrasound-assisted extraction of phenolic compounds from *Psidium guajava* L. leaves. *Food Anal. Met.* **2017**, *10*, 2781–2791. [CrossRef]

27. Carciochi, R.A.; Sologubik, C.A.; Fernández, M.B.; Manrique, G.D.; D'Alessandro, L.G. Extraction of antioxidant phenolic compounds from brewer's spent grain: Optimization and kinetics modeling. *Antioxidants* **2018**, *7*, 45. [CrossRef] [PubMed]

28. Robbins, R.J.; Leonczak, J.; Johnson, J.C.; Li, J.; Kwik-Uribe, C.; Prior, R.L.; Gu, L. Method performance and multi-laboratory assessment of a normal phase high pressure liquid chromatography-fluorescence detection method for the quantitation of flavanols and procyanidins in cocoa and chocolate containing samples. *J. Chromatogr. A* **2009**, *1216*, 4831–4840. [CrossRef] [PubMed]

29. Moreira, M.M.; Morais, S.; Carvalho, D.O.; Barros, A.A.; Delerue-Matos, C.; Guido, L.F. Brewer's spent grain from different types of malt: Evaluation of the antioxidant activity and identification of the major phenolic compounds. *Food Res. Int.* **2013**, *54*, 382–388. [CrossRef]

30. Ikram, S.; Huang, L.; Zhang, H.; Wang, J.; Yin, M. Composition and nutrient value proposition of brewers spent grain. *J. Food Sci.* **2017**, *82*, 2232–2242. [CrossRef]

Article

Intensification of Polyphenol Extraction from Olive Leaves Using Ired-Irrad®, an Environmentally-Friendly Innovative Technology

Anna-Maria Abi-Khattar [1], **Hiba N. Rajha** [1], **Roula M. Abdel-Massih** [2], **Richard G. Maroun** [1,*], **Nicolas Louka** [1] and **Espérance Debs** [2]

1   Centre d'Analyses et de Recherche, Unité de Recherche Technologies et Valorisation Agro-alimentaire, Faculté des Sciences, Université Saint-Joseph, B.P. 17-5208 Riad El Solh, Beirut 1104 2020, Lebanon
2   Department of Biology, University of Balamand, Tripoli P. O. Box 100, Lebanon
*   Correspondence: richard.maroun@usj.edu.lb

Received: 27 June 2019; Accepted: 13 July 2019; Published: 18 July 2019

Abstract: Optimization of infrared-assisted extraction was conducted using Response Surface Methodology (RSM) in order to intensify polyphenol recovery from olive leaves. The extraction efficiency using Ired-Irrad®, a newly-patented infrared apparatus (IR), was compared to water bath (WB) conventional extraction. Under optimal conditions, as suggested by the model and confirmed experimentally, the total phenolic content yield was enhanced by more than 30% using IR as contrasted to WB, which even required 27% more ethanol consumption. High Performance Liquid Chromatography analyses quantified the two major phenolic compounds of the leaves: Oleuropein and hydroxytyrosol, which were both intensified by 18% and 21%, respectively. IR extracts increased the antiradical activity by 25% and the antioxidant capacity by 51% compared to WB extracts. On the other hand, extracts of olive leaves obtained by both techniques exhibited equal effects regarding the inhibition of 20 strains of *Staphylococcus aureus*, with a minimum inhibitory concentration (MIC) varying between 3.125 and 12.5 mg/mL. Similarly, both extracts inhibited Aflatoxin B1 (AFB1) secretion by *Aspergillus flavus*, with no growth inhibition of the fungus. Finally, optimization using RSM allowed us to suggest other IR operating conditions aiming at significantly reducing the consumption of energy and solvent, while maintaining similar quantity and quality of phenolic compounds as what is optimally obtained using WB.

Keywords: olive leaves; infrared-assisted extraction; response surface methodology; antioxidants; antimicrobial activity

1. Introduction

The cultivation of olive trees is a widespread practice in the Mediterranean region, accounting for about 98% of the world's olive cultivation [1]. The olive tree is gradually expanding, with nearly 18 Mt of olives harvested yearly around the world. Olive tree culture and the olive processing industry produce large amounts of byproducts. It has been estimated that pruning alone produces 25 kg of byproducts (twigs and leaves) per tree, annually [2]. Olive leaves are considered to be an easily available agricultural byproduct [3,4]. They represent around 10% of the total weight of olives upon harvesting [5,6]. Olive leaves are a very rich source of bioactive compounds such as secoiridoids, flavonoids, and triterpenes [7]. They can potentially have a higher added value if their fate is reconsidered.

Valorization of the residual biomass derived from the agricultural and food sector is nowadays regarded as central to the emerging bioeconomy. This biomass is definitely underrated, despite its richness in valuable substances [8]. Olive leaves are usually disposed as waste. Otherwise, their infusion can be used in folk medicine [9]. The secoiridoid oleuropein is the main compound, along with other secoiridoids derived from tyrosol and flavonoids [5,6,9–11]. Other olive byproducts such as olive mill waste and mill wastewater, and wet olive pomace, have also been investigated for their polyphenols content [12].

Phenolic compounds, plant secondary metabolites, are gaining more and more interest in the agro-industrial sector. They are being extensively studied, majorly due to their biological effects. Therefore, they are the subject of numerous extraction techniques used to recover them out of their original matrices. In this regard, conventional extraction using organic solvents is the most widely used method. Nevertheless, environmental toxicity, long duration of processing, and consumption of large quantities of organic solvents are the major concerns arising from this method [13]. These major drawbacks led the researchers to seek new technologies to be applied or to be combined with pre-existing ones.

Many extraction technologies were used for the intensification of polyphenols recovery from plant materials, such as ultrasound assisted extraction [14,15], microwave assisted extraction [16,17], pressurized liquid assisted extraction [18], supercritical fluid extraction [19,20], and others. Optimization of any given extraction technique goes obviously through a maximization of polyphenols recovery while maintaining their chemical integrity and, subsequently, their functional activities.

Infrared irradiation is one of these alternatives introduced as a 'green' energy source [21,22] used to boost the extraction of natural products. The infrared assisted extraction apparatus Ired-Irrad® is a new generation of ecofriendly machines that enhance the extraction of bioactive compounds from natural matrices using a ceramic infrared emitter [22]. Recently, this technique has been explored on pomegranate peels [14], *Prunus armeniaca* L. pomace [23], apricot pomace [24,25], and *Saussurea lappa* [26], and permitted the intensification of polyphenol recovery compared to conventional extraction methods. Infrared-assisted extraction is easy to use, economical, requires low energy consumption [23], and has a great potential to be scaled-up to an industrial level.

The innovation of this study relies on the use of a new patented technique based on infrared apparatus (IR) irradiations for the recovery of bioactive compounds, while preserving their biological properties. To our knowledge, no previous studies have investigated this effect on olive leaves. Our ultimate aim is to optimize extraction of total phenolic content from olive leaves using IR irradiation, and to compare the results with the ones obtained using water bath conventional extraction. Moreover, quality of both extracts will be inspected by testing their antioxidative and antiradical activities, their antibacterial effect against 20 strains of *Staphylococcus aureus* and 7 strains of *Escherichia coli*, and their antifungal effect not only against *Aspergillus flavus* growth, but also against its production of aflatoxin B1.

2. Materials and Methods

2.1. Plant Material

Olive leaves were provided by a local olive mill in northern Lebanon El Koura in September 2018. The leaves were washed with water to remove impurities such as dust, then dehydrated in an airflow oven at 40 °C for 48 h. Dried leaves were ground (Philips, United Arab Emirates, MEA) and then sieved using a vibrating multi sieve separator (ELE International, Loveland, CO, USA). Ground leaves, from 0.85 to 2 mm in size, were packed in plastic bags and stored at ambient temperature in the dark for further use.

2.1.1. Dry Matter

Initial and final moisture contents were determined by drying the leaves for 24 h in a ventilated oven at 105 °C. The dry matter (DM) of raw material was 91 ± 0.4%.

2.1.2. Chemicals

All chemicals used in the experiments were analytical grade. Folin-Ciocalteu reagent, sodium carbonate, gallic acid, 1,1-diphenyl-2-picrylhydrazyl (DPPH), 6-hydroxy-2,5,7,8-tetramethylchroman-2-carboxylic acid (Trolox), ascorbic acid, sulfuric acid, sodium phosphate, ammonium molybdate, oleuropein, and hydroxytyrosol were purchased from Sigma-Aldrich, Darmstadt, Germany.

2.2. Experimental Methods

2.2.1. Water Bath Extraction

The conventional extraction was carried out in a digital water bath (JSR JSWB-22T, Gongju-city, Korea) (Figure 1a).

2.2.2. Infrared-Assisted Extraction

The infrared-assisted extraction apparatus (Ired-Irrad®, Beirut, Lebanon) was designed and patented in collaboration between the faculties of Sciences of Saint-Joseph University (Beirut, city, Lebanon) and the University of Balamand (Kelhat, city Lebanon) [22]. The extraction prototype consists of a ceramic infrared emitter, linked to a proportional-integral-derivative (PID) control and temperature adjustment system. The sample consisting of olive leaves and solvent was placed in a round bottom flask connected to a condenser at a 1 cm distance from the ceramic IR emitter (Rotfil, Pianezza, Italy) (Figure 1b).

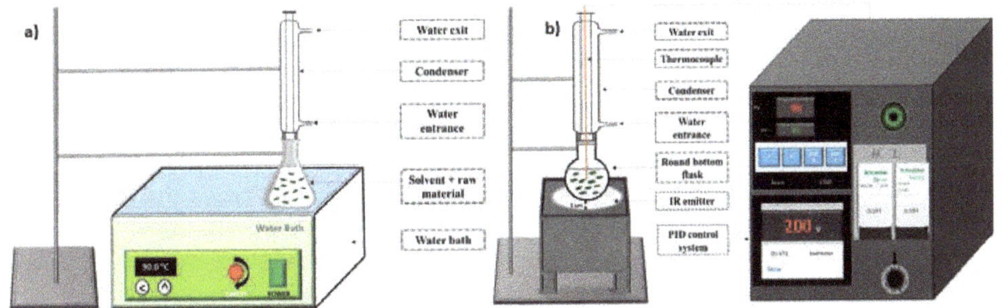

Figure 1. Instrumental setup for (**a**) water bath (WB) and (**b**) infrared apparatus (IR).

2.2.3. Extraction Procedure

An amount of 5 g of ground leaves was added to 100 mL of solvent consisting of varying amounts of aqueous ethanol. Extractions were carried out at predetermined temperatures and time periods. The fixed particle size (0.85–2 mm) and solid to liquid ratio (1:20 *w/v*) were chosen based on a preliminary set of experiments (Figure 2a,b). Once the extraction was complete, the extracts were filtered through glass wool, then centrifuged for 10 min at 4500 rpm and stored at −20 °C until analyses. Prior to HPLC analyses, the supernatants were filtered using a 0.45 μm syringe after centrifugation [27].

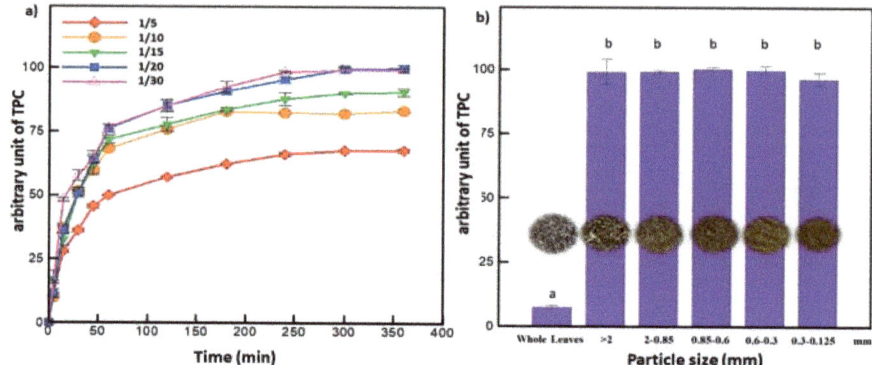

Figure 2. (**a**) Effect of solid to liquid ratio and (**b**) particle size on the extraction yield (letters a and b indicate significant statistical difference between means). Every arbitrary unit value is the ratio of the Total Phenolic Compounds (TPC) (at the corresponding experimental conditions) to the highest obtained TPC.

2.2.4. Total Phenolic Compounds

The total phenolic content was determined according to the Folin-Ciocalteu method [28,29]: 0.2 mL of each extract were mixed with 1 mL of ten-fold diluted Folin–Ciocalteu reagent (Sigma-Aldrich, Darmstadt, Germany), and 0.8 mL of sodium carbonate (Na_2CO_3) (75 g/L) (Sigma-Aldrich, Darmstadt, Germany) were added to the mixture. The absorbance was then measured by a UV-Vis spectrophotometer (Biochrom Ltd., Cambridge, England) at 750 nm. The total phenolic content was expressed as mg of Gallic Acid Equivalents per gram of dry matter mg Gallic Acid Equivalent/g DM.

2.3. Experimental Design

Response surface methodology (RSM) is an assemblage of statistical and mathematical methods used for products developing, improving, and optimizing processes [30]. It permits to measure the linear and quadratic effects of parameters, as well as the probable interactions between the variables.

Optimization of phenolic compounds extraction from ground olive leaves was carried out using RSM. A rotatable central composite design (2^3 + star) (22 runs: 8 factorial design points, 6 star points and 8 center points with 12 degrees of freedom) was created to evaluate the main impact of three experimental factors: Solvent mixture, time, and temperature on the response parameter: Total Phenolic Compounds (TPC). The same design was applied twice: (1) For the extraction process using the conventional water bath (WB), and (2) for the infrared-assisted (IR) extraction apparatus. Ethanol percentage values varied between 40% and 80%, time between 60 and 180 min, and temperature between 38 °C and 77 °C (considered as −1 and +1 levels, respectively). Solvent mixture, time, and temperature are independent variables that were coded at five levels (−α, −1, 0, +1, +α). Considering three parameters and one response, experimental data were fitted to obtain a second-degree regression equation of the form:

$$Y = \beta_1 + \beta_2 E + \beta_3 t + \beta_4 T + \beta_5 E^2 + \beta_6 t^2 + \beta_7 T^2 + \beta_8 Et + \beta_9 ET + \beta_{10} tT \tag{1}$$

where Y is the predicted response parameter; β_1 is the mean value of responses at the central point of the experiment; β_2, β_3 and β_4 are the linear coefficients; β_5, β_6 and β_7 are the quadratic coefficients; β_8, β_9 and β_{10} are the interaction coefficients; E is the solvent mixture; t is the extraction time; and T is the extraction temperature. Experimental design and statistical treatment of the results were performed using STATGRAPHICS Centurion XVII (Statgraphics 18, The Plains, Virginia).

2.4. High Performance Liquid Chromatography

Polyphenol (oleuropein and hydroxytyrosol) identification and quantification were conducted by HPLC, using an HPLC-DAD (diode array detection) (Waters Alliance, Milford, MA, USA), a quaternary Waters e2695 pump, an UV–vis photodiode array spectrophotometer (Waters Corporation, Milford, USA), a control system, and a data collection Empower 3 software. Analyses were carried out on a Discovery HS C18, 5 µm, 250 × 4.6 mm, column (Supelco, Bellefonte, PA, USA) with a HS C18, Supelguard Discovery, 20 × 4 mm, 5 µm, precolumn (Supelco, Bellefonte, PA, USA). The column temperature was maintained at 25 °C. Separation of 10 µL was performed at a flow rate of 0.8 mL min^{-1}. Mobile phase A consisting of 0.5% (*v/v*) acetic acid in water and mobile phase B consisting of 100% acetonitrile were used. Solvent gradient changed according to the following conditions: From 0 to 10 min, 95% (A): 5% (B) to 70% (A): 30% (B); from 10 to 12 min, 70% (A): 30% (B) to 67% (A): 33% (B); from 12 to 17 min, 67% (A): 33% (B) to 62% (A): 38% (B); from 17 to 20 min, 62% (A): 38% (B) to 50% (A): 50% (B); from 20 to 23 min, 50% (A): 50% (B) to 5% (A): 95% (B); from 23 to 25 min, 5% (A): 95% (B) to 95% (A): 5% (B); from 25 to 35 min, 95% (A): 5% (B) to 95% (A): 5% (B). Spectrophotometric detection wavelength was carried out at 280 nm. Identification of the compounds was based on retention time of standards and comparison of spectra [31].

2.5. Antioxidant Activity

The total antioxidant activity of the extracts was determined using the phosphomolybdenum reduction essay [32]. The principle of this method is the formation of a green complex phosphate Mo (V). A quantity of 100 µL of each extract were mixed with 1 mL of the reagent solution (0.6 M sulfuric acid, 28 mM sodium phosphate, and 4 mM ammonium molybdate). Samples were incubated for 90 min at 95 °C. Absorbance was then measured at 695 nm. Antioxidant activity was expressed as µg of Ascorbic Acid Equivalent per milliliter (µg AAE/mL).

2.6. Antiradical Activity

Free radical scavenging activity was measured by the capacity of the phenolic compounds to reduce DPPH (2,2-diphenyl-picrylhydrazyl), according to [33]. 1.45 mL of DPPH (0.06 mM) (Sigma-Aldrich, St-Quentin Fallavier, France) radical was added to 50 µL of olive leaf extracts or Trolox (positive control) (Sigma-Aldrich, St-Quentin Fallavier, France). After 30 min of incubation at room temperature in the dark, the absorbance was measured at 515 nm using pure methanol as a blank. The inhibition percentage of the DPPH free radical is calculated as follows: Inhibition Percentage = [(absorbance of negative control − absorbance of sample)/absorbance of negative control] × 100. Antiradical activity was expressed as µg of Trolox Equivalent per milliliter (µg TE/mL) [34].

2.7. Antifungal Activity

Aspergillus flavus NRRL (Northern Regional Research Laboratory) 62477 isolates from spices were grown in Petri dishes containing malt extract agar (MEA) at pH 5.5 ± 0.3 for 7 days at 27 °C. A spore suspension was then prepared using Tween 80 solution. The spores were counted on a Neubauer haemocytometer (Superior, Marienfeld, Lauda-Konigshofen, Germany). The final concentration of the spore suspension was adjusted to 10^5 spores/mL.

2.8. Fungal Growth Inhibition

Olive leaf extracts (125 and 250 µg) were added to the MEA medium. A final volume of 20 mL of MEA was transferred in petri dishes. For the control culture, 20 mL of MEA without polyphenols were poured in a petri dish. Afterwards, 10 µL of the previously prepared spore solution (10^5 spores/mL) were placed in the center of each petri dish. All the dishes were left for 7 days in the incubator at

27 °C. On the seventh day, the diameters of all the cultures were measured. The *A. flavus* inhibition percentage was calculated as follows:

$$\text{Inhibition \%} = [(\text{Initial diameter} - \text{diameter after incubation})/\text{initial diameter}] \times 100 \qquad (2)$$

2.9. Aflatoxin B1 (AFB1) Inhibition

Aflatoxin B1 (AFB1) inhibition was detected using reverse phase HPLC (diode array detection) (Waters Alliance, Milford, MA, USA) coupled with a fluorescence detector and a C18 column 5 μm, 250 × 4.6 mm, column (Supleco, Bellefonte, PA, USA) fitted with a HS C18, Supelguard Discovery, 20 × 4 mm, 5 μm, precolumn (Supelco, Bellefonte, PA, USA). The column temperature was maintained at 40 °C. The mobile phase was composed of HPLC water: Methanol: Nitric acid 4M (55:45:0.35 *v/v/v*) and 119 mg/L KBr prepared and filtered on the same day of HPLC analysis, performed with a flow rate of 0.8 mL/min. The injection volume was 100 μL, the cycle duration was 35 min, and the wavelengths for excitation and emission were 360 and 430 nm, respectively.

2.10. Antibacterial Activity

2.10.1. Microorganisms Used

Twenty bacterial strains (American Type Culture Collections ATCC, Newman and clinical strains) of Gram-positive *S. aureus* and seven strains of Gram-negative *E. coli* (one *E. coli* 25921 DSM 1103 and the others are strains of different profiles of resistance) that were isolated from patients at the Centre Hospitalier Du Nord Hospital (CHN, Zghorta, Lebanon), were used in this study.

2.10.2. Determination of Minimal Inhibitory Concentration for Extracts

The macro-dilution broth method was used for the determination of the minimum inhibitory concentration (MIC) of *Olea europea* extracts, as described by the Clinical and Laboratory Standards Institute [35]. A standardized bacterial inoculum was prepared and adjusted to 0.5 McFarland, then diluted to 10^6 CFU/mL. Leaf lyophilized extracts were diluted with DMSO to produce two-fold serial dilutions ranging from 0.39 to 50 mg/mL. 1 mL of broth was added to each tube of the macro-dilution tray. 300 μL of plant extract suspension were added to the first tube in each series, after removing the same volume of broth, in order to achieve the final desired concentration. 1 mL of bacterial inoculum was added to each tube to reach 2 mL of final volume. The final extract concentration in each tube is presented in Table 1. Broth (2 mL) was used as a negative control, whereas 1 mL of Mueller-Hinton broth and 1 mL bacterial suspension were used as a positive control. The tray was then incubated for 24 h at 35 °C. Thereafter, the test tubes were checked for turbidity and MIC was determined by observing the lowest concentration of extract where there is no visible bacterial growth compared to the negative and positive control. The antibacterial analyses were repeated twice and gave the same MIC values.

Table 1. Minimum Inhibitory Concentrations (MICs) of WB and IR olive leaves extract against twenty *Staphylococcus aureus* strains.

Bacterial Strain	Olive Leaves Extract Minimum Inhibitory Concentrations (mg/mL)	
	WB	IR
S. aureus 001	12.5	12.5
S. aureus 002	6.25	6.25
S. aureus 003	12.5	12.5
S. aureus 004	3.125	3.125
S. aureus 005	3.125	3.125
S. aureus 006	12.5	12.5
S. aureus 007	12.5	12.5

Table 1. *Cont.*

Bacterial Strain	Olive Leaves Extract Minimum Inhibitory Concentrations (mg/mL)	
	WB	IR
S. aureus 008	12.5	12.5
S. aureus 009	12.5	12.5
S. aureus 010	12.5	12.5
S. aureus 011	12.5	12.5
S. aureus 012	12.5	12.5
S. aureus 013	12.5	12.5
S. aureus 014	12.5	12.5
S. aureus 015	12.5	12.5
S. aureus 016	12.5	12.5
S. aureus 017	12.5	12.5
S. aureus Newman	6.25	6.25
S. aureus ATCC 29213	6.25	6.25
S. aureus N315 MRSA	12.5	12.5

2.11. Statistical Analysis

All experiments and measurements were done in triplicates. The mean values and the standard deviations were calculated. Error bars, in all figures, correspond to the confidence level 95%. Variance analyses (ANOVA) and Least Significant Difference (LSD) tests were done by STATGRAPHICS® Centurion XV (Statgraphics 18, The Plains, Virginia).

3. Results and Discussion

3.1. Solid to Liquid Ratio and Particle Size Selection

The selection of the solid to liquid ratio and the particle size was based on primary studies (Figure 2a,b). The extraction yield increased by 50% while decreasing the solid to liquid ratio form 1:5 to 1:20 (g/mL). For a higher liquid ratio (1:30), the TPC remained constant. Hence, the solid to liquid ratio of 1:20 g/mL was adopted for all the subsequent experiments. On the other hand, lowering the particle size of olive leaves from whole leaves to 2, 0.85, 0.6, and 0.3 mm increased the contact surface area between the samples and the solvent, thus leading to a higher extraction yield. The grinding process up to 0.3 mm gave around 14 times higher recovery of polyphenols compared to whole leaves. However, for the particle size below 0.3 mm, the extraction yield decreased. Similar results were found in the literature where olive leaves of particle sizes below 0.2 mm decreased the extraction yield of polyphenols [36]. Too fine particles tend to agglomerate, limiting the accessibility of the solvent to the solid matrix and consequently altering the mass transfer phenomenon [37]. For that reason, 0.85 to 2 mm particle size was selected to be used in the following experiments.

3.2. Effect of Solvent Mixture, Time, and Temperature on TPC Extraction

The response surface methodology was conducted in the aim of studying and determining the optimal conditions for the highest phenolic concentration in WB and IR extracts. Based on the abovementioned results, a central composite design was conducted using RSM with the selected particle size (0.85–2 mm) and solid to liquid ratio (1:20 (g/mL)), varying solvent mixture, time, and temperature (Table 2). Values varied between 14 and 25 mg/g DM for WB and between 15 and 33 mg/g DM for IR leading to a higher TPC range with IR treatment.

Table 2. Arrangement for independent variables and their responses for TPC (mg GAE/g dry matter (DM)).

Run	Central Composite Design	Variable Levels Uncoded			Phenolic Compounds Yield (mg GAE/g DM)			
		Solvent (% Ethanol)	Time (min)	Temperature (°C)	WB		IR	
					Experimental	Predicted	Experimental	Predicted
1		40	60	38	15.74	15.00	19.71	18.27
2		80	60	38	14.41	13.81	16.61	14.73
3		40	180	38	18.35	18.47	23.20	22.40
4	Factorial design	80	180	38	16.12	16.65	18.39	16.63
5	points	40	60	77	22.52	21.16	21.02	22.15
6		80	60	77	23.90	22.94	23.66	23.81
7		40	180	77	23.81	23.58	27.51	28.75
8		80	180	77	24.81	24.72	27.38	28.18
9		26.36	120	57.5	17.41	18.33	21.90	21.51
10		93.63	120	57.5	18.04	18.30	16.76	18.06
11	Star points	60	19.09	57.5	16.31	18.09	16.46	17.37
12		60	220.9	57.5	23.11	22.51	23.86	23.86
13		60	120	24.7	14.31	14.31	15.01	18.20
14		60	120	90.3	25.11	26.28	33.47	31.18
15		60	120	57.5	21.20	21.22	21.77	20.81
16		60	120	57.5	21.15	21.22	20.17	20.81
17		60	120	57.5	21.43	21.22	20.32	20.81
18		60	120	57.5	21.09	21.22	20.70	20.81
19	Center points	60	120	57.5	21.20	21.22	20.45	20.81
20		60	120	57.5	21.49	21.22	20.51	20.81
21		60	120	57.5	20.47	21.22	21.29	20.81
22		60	120	57.5	21.89	21.22	21.45	20.81

The impact of the three studied parameters on the corresponding response (TPC) was analyzed according to the Pareto chart and 3D-mesh for both WB and IR (Figure 3).

Temperature and time exhibited significant linear positive effects on polyphenol extraction from olive leaves using both WB and IR ($p < 0.01$) (Figure 3a,b). Temperature rise led to a TPC increase to reach optimum at 90 °C. In addition, ethanol concentration displayed a highly significant linear negative effect in case of IR ($p < 0.01$). This might be due to the adequation existing between the used IR wavelengths and the absorption characteristics of water molecules. In this sense, the increase in ethanol percentage negatively affected the IR extraction efficiency.

Temperature had a positive quadratic effect on TPC recovery by IR. Nonetheless, a combination of linear and quadratic positive effects suggests the presence of a latency phase followed by a fast increase in the variation of TPC as a function of temperature. For WB extraction, and despite that ethanol displayed a non-significant effect, it exhibited a quadratic significant negative effect on TPC (Figure 3a,b). This observation implies that the majority of the extracted phenolic compounds of olive leaves were recovered with an intermediate solvent polarity varying between that of pure water and pure ethanol. Previous studies discussed the interaction between the IR wavelength emitted and its capacity of being absorbed by the solvent, which is in relation with the solvent polarity. The elemental and most important features of infrared radiation are the high heat transfer capacity, heat penetration straight into the product, fast management response, and good chances for process control [38].

Ethanol time interaction had a positive effect ($p < 0.05$) on TPC, with a higher impact observed using IR (Figure 3b). Although ethanol played a negative linear effect on TPC, it positively amplified the positive linear effect of temperature. Other mildly significant interactions exist in the case of IR.

Finally, 3D mesh diagrams showed the spectrum of phenolic compounds' yields simultaneously as a function of solvent mixture, time, and temperature (Figure 3c,d). The red areas highlight the ranges of the highest TPC reached with different combinations of the three operating conditions.

Increasing both time and temperature led to an increase in the TPC in both WB and IR extracts, giving a maximum recovery at 90 °C. Diffusion coefficient is well known to be directly proportional to temperature elevation [39]. Furthermore, increasing the extraction time at a constant temperature also appeared to heighten TPC. This phenomenon is explained by Fick's second law of diffusion that predicts a final equilibrium between the solute concentrations in the solid matrix and in the extraction solution after a certain time [40].

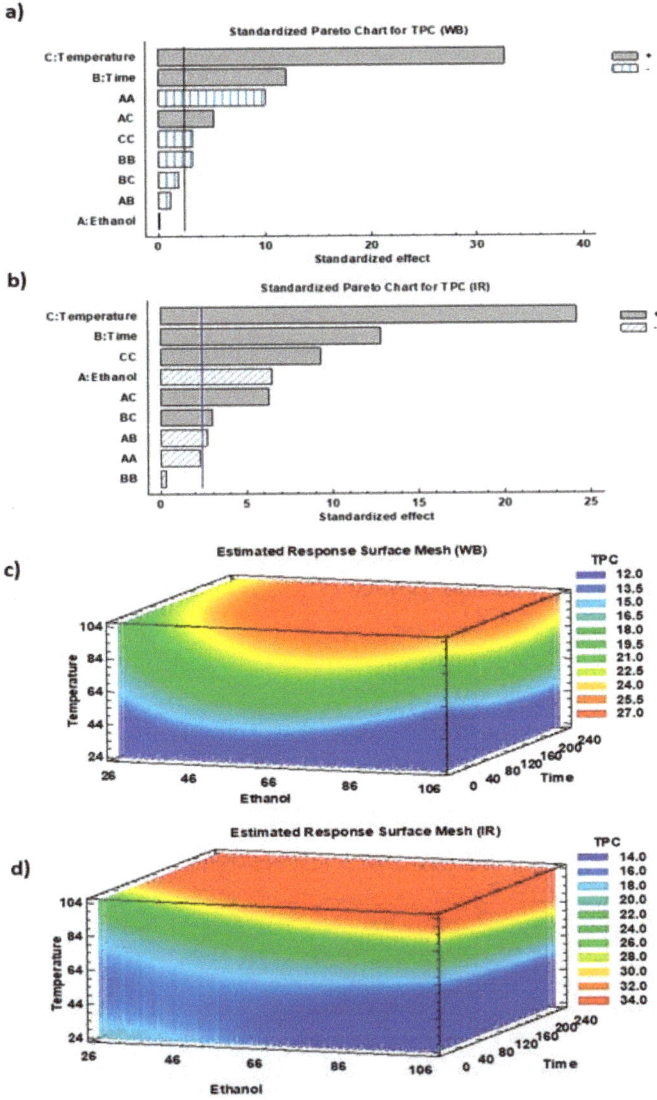

Figure 3. (**a,b**) Standardized Pareto charts and (**c,d**) the corresponding estimated response surface mesh of TPC for WB and IR.

3.3. Optimization of TPC Extraction

WB optimum TPC yield was reached with 70% ethanol/water, 193 min, and 90 °C, while IR optimum was reached with 55% ethanol/water, 220 min, and 90 °C (Table 3). Maximum extraction yields for the optimum conditions were 27.12 and 36.23 mg GAE/g DM for WB and IR, respectively. These suggested optimums reflect the operating conditions required for the recovery of the highest polyphenol content in both WB and IR.

Table 3. Optimum extraction conditions for WB and IR.

Factor	Optimum Conditions	
	WB	**IR**
Ethanol/Water (%)	70.16	55.35
Time (min)	193.28	220.91
Temperature ($^\circ$C)	90.29	90.29

Response values were given by statistical analysis to fit the model equation (Table 4). The regression equation allowed for calculating the TPC predicted values that will be lately compared to the experimental ones.

Table 4. Second order regression equation for TPC of each extraction technique and the R-squared of each equation.

Extraction Technique	R^2 (Percent)	Equation
WB	90.69	TPC = 29.1281 − 0.0854381E + 0.0347351t − 0.487806T − 0.000850084E^2 − 0.0000128046t^2 + 0.00366906T^2 − 0.00046598Et + 0.00333862ET + 0.000528357tT
IR	95.23	TPC = −1.18369 + 0.214169E + 0.0644087t + 0.193729T − 0.0025667E^2 − 0.0000897562t^2 − 0.000854751T^2 − 0.000132978Et + 0.0019014ET − 0.000225784tT

In order to verify the predictive capacity of the model, optimum conditions were revalidated and obtained values (26.31[a] ± 0.8 and 34.28[b] ± 1.9 mg/g DM) were compared to the values predicted by the model (Table 5). Predicted values displayed a great correlation with experimental values and resulted in the calculation of the coefficients of determination (R^2) which are 95% and 90% for WB and IR respectively.

Table 5. Predicted and experimental results of TPC (mg GAE/g DM) for WB and IR.

Optimum TPC Value (mg GAE/g DM)	Predicted	Experimental
WB	27.12	26.31[a] ± 0.3
IR	36.23	34.28[b] ± 1

a and b indicate significant statistical difference between means.

A comparison between both optima was the main objective of this response study. The maximum amounts of TPC were obtained at an ethanol concentration of 70% and 55% for WB and IR, respectively. This is in accordance with previous investigations, which indicated that the conventional extraction of biophenols from olive leaves with water/ethanol solutions peaked at 70% ethanol [27]. The recovery of polyphenols from olive leaves was also optimized by pressurized liquid extraction. The optimal conditions were found to be: 50% ethanol/water mixture at 80 °C, with 2 cycles of 5 min giving a TPC of 53.15 mg GAE/g [41]. Furthermore, 80% ethanol [42] and 60% ethanol [43] gave the highest yield using ultrasound-assisted extraction in precedent studies. Compared to these ultrasound extractions, IR permitted the reduction of ethanol consumption, which is of great economic interest when extended to an industrial level. In addition, reducing the use of organic solvent consumption limits its related environmental damage and is considered to be one of the most important principles of green chemistry. Organic solvents react in the atmosphere under sunlight, producing air pollutants that seriously affect human, animal, and plant health. To this end, a very important future challenge resides in the development of new environmentally-friendly analytical methods capable of giving good quality results [44].

Many studies evaluated the phenolic composition of the enhanced recovery of olive leaf extracts by different technologies. Analogous amounts (24.36 ± 0.85 mg GAE/g DM) were reached with conventional extraction in a previous study using the same ethanol/water concentration [45]. Obtained results were also comparable to those found earlier for similar extracts, recovered with ultrasound-assisted extraction (25.06 mg GAE/g DM [46], 38.66 mg GAE/g DM [47]). In another study, ultrasound showed an improvement in the extraction yield compared to conventional extraction [48]. Also, resembling yield (37.52 ± 0.87 mg GAE/g DM) was attained using subcritical water extraction but while using 56% higher temperature and 60% more solvent usage. Likewise, microwave-assisted extraction was studied and compared to conventional method: 55% extraction yield amelioration compared to WB using 80% methanol/water [31].

The suggested IR optimum of TPC obtained by the RSM study can be further reconsidered in order to fit different industrial or environmental demands. For example, in order to reach with IR technology, the same optimal TPC (27.12 mg GAE/g DM) acquired by WB, ethanol could be reduced to 12%, temperature to 40 °C, and time raised to 240 min.

In the subsequent sections, extractions were carried out in the optimum conditions using WB and IR and used in the upcoming analyses.

3.4. High Performance Liquid Chromatography

It has been frequently reported that oleuropein and hydroxytyrosol are considered as major phenolic compounds of olive leaves [49,50]. HPLC analyses were done for both WB and IR optimum extracts in order to compare oleuropein and hydroxytyrosol contents, and then to correlate these results with the TPC values.

As shown in Table 6, optimum conditions led to highest values of oleuropein and hydroxytyrosol contents in IR extract of 14.01^b ± 0.9 and 0.40^d ± 0.008 mg/g DM, while 11.84^a ± 1.2 and 0.33^c ± 0.02 mg/g DM in WB extract, respectively. In a previous investigation, it was reported that the oleuropein yield was 6.53 ± 0.01 mg GAE/g DM (2.15 times lower than our results), and hydroxytyrosol 0.54 ± 0.02 mg GAE/g DM (almost the same amount as our findings), following an ultrasonic extraction for 3 min at 50 °C using 75% ethanol/water [51]. These valuable compounds in olive leaf extracts are responsible for many health benefits. Oleuropein is reported to have antimicrobial activities against viruses, retroviruses, bacteria, yeasts, fungi, and other parasites [52]. On the other hand, hydroxytyrosol is known to be beneficial for treating atherosclerosis and prevents diabetic neuropathy [53]. In particular, both molecules have demonstrated high antioxidant and antimicrobial activities [54]. Therefore, there is a growing interest to use these bioactive molecules in various industrial applications in food supplements in the pharmaceutical and cosmetic industries [55]. Antioxidant, antiradical, antibacterial, and antifungal activities of the two optimum extracts were thus compared.

Table 6. Oleuropein and hydroxytyrosol concentrations (mg/g DM) in WB and IR olive leaf extracts.

Concentration (mg/g DM)	Extraction Technique	
	WB	IR
Oleuropein	11.84^a ± 1.2	14.01^b ± 0.9
Hydroxytyrosol	0.33^c ± 0.02	0.40^d ± 0.008

Different letters (a, b, c, and d) indicate significant statistical difference between means.

3.5. Antioxidant and Antiradical Activity

Two methods were used to assess the antioxidant and antiradical capacities of olive leaf extracts: The phosphomolybdenum and the DPPH assays (Figure 4a).

IR extract gave the greatest antioxidant activity 4002.94^a ± 24 µg AAE/mL compared to WB 2653.23^b ± 263 µg AAE/mL (Figure 4a). The IR extract also exhibited the highest antiradical activity 3237^a ± 115 µg TE/mL compared to that of WB: 2589^b ± 93 µg TE/mL. A previous study displayed a lower antiradical activity of about 1930 ± 0.09 µg TE/mL, using ultrasound-assisted extraction of polyphenols from olive leaves at 50 °C with 75% ethanol/water [51].

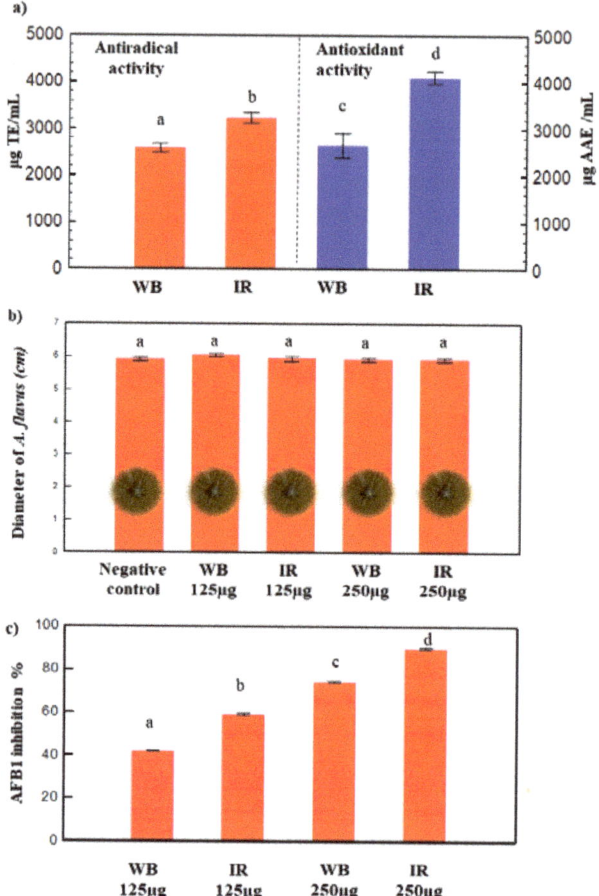

Figure 4. (**a**) Antiradical and antioxidant activities, (**b**) fungal growth of *A. flavus* and (**c**) AFB1 inhibition percentage for WB and IR extracts (a, b, c, and d indicate significant statistical difference between means).

3.6. Antifungal Activity

The phenolics obtained by the optimal extraction conditions of WB and IR did not affect the fungal growth, but significantly inhibited the AFB1 production by *A. flavus* (Figure 4b,c). The observed inhibition percentages of the toxin varied between 40% and 90%, with the increase of the polyphenol concentration of the extract from 125 µg to 250 µg (Figure 4c). Both WB and IR extracts were able to disrupt the AFB1 production by *A. flavus*, without having any significant effect on the fungal growth (Figure 4b,c). However, the impact of IR on the inhibition of aflatoxin B1 biosynthesis was ≈1.3 times higher compared to WB. This reflects the synergetic effect of phenolic compounds present in each extract that led to a modified biological efficiency for WB and IR, even at the same polyphenol concentration. The synergistic effects of phenolics against the growth of *A. flavus* using IR was also observed for pomegranate peel extracts [14]. AFB1 is known to be the most predominant and toxic aflatoxin. It is also known as being one of the mainly genotoxic agents and hepatocarcinogens identified [56]. It is very interesting to be able to inhibit the mycotoxin secretion without affecting the fungal growth and modifying the microbial ecology of the crop.

3.7. Antibacterial Activity

Different studies claim strong antimicrobial activity of oleuropein and hydroxytyrosol against many bacteria (e.g. *Bacillus cereus, Listeria monocytogenes*, etc.) [17,57,58]. In the present study, WB and IR olive leaf extracts showed antibacterial activity against all the tested *S. aureus* strains, with mean MIC values that range between 3.125 and 12.5 mg/mL (Table 1). The lowest MIC was recorded for *S. aureus 004* and *S. aureus 005* as 3.125 mg/mL. Both WB and IR extracts exhibited similar antibacterial activity suggesting that active compound(s) were not affected by the IR extraction method. Moreover, the IR technology ameliorated the antioxidant and antiradical activities, as well as it inhibited the AFB1 production (Figure 4).

For all the tested concentrations of polyphenols (0.39; 0.78; 1.56; 3.12; 6.25; 12.5; 25; and 50 mg/mL), no antibacterial activity against *E. coli* strains was detected.

Similarly, olive leaf extracts obtained with mechanical stirring for 12 hours, using acetone, did not show any antibacterial activity against *E. coli* (ATCC 25922); however, an MIC of 2.5 mg/mL against *S. aureus* (ATCC 25923) was observed [59]. This might be due to the fact that Gram-negative bacteria possess an outer membrane that acts as a barrier to many environmental substances [60].

4. Conclusions

This study revealed the efficiency of Ired-Irrad® technology for the intensification of polyphenol recovery from olive leaves. Time and temperature were shown, by response surface methodology, to be the most significantly-affecting infrared-assisted extraction parameters. Compared to the conventional methods, IR allowed the extraction of a higher yield of polyphenols and improved many of their biological activities—i.e., antioxidant, antiradical, and anti-AFB1 secretion. Compared to WB, IR technology enhanced the recovery of both oleuropein and hydroxytyrosol, the two main polyphenols present in olives leaves. IR seems to be a very promising new generation of ecofriendly machines that can enhance the extraction of polyphenols with less energetic and solvent consumptions.

5. Patents

Rajha, H.N., Debs, E., Maroun, R.G., Louka, N. (2017). System for extracting, separating or treating products through infrared radiation. Adequacy between the properties of infrared radiation and those of the processed products. Invention patent number 2017 / 11-11296L granted on 29/11/2017.

Author Contributions: Conceptualization, methodology, writing—review and editing, E.D., H.N.R., N.L., R.M.A.M., and R.G.M.; investigation, visualization and writing—original draft preparation, A.M.A.K.

Funding: This project has been jointly funded with the support of the National Council for Scientific Research of Lebanon CNRS-L and University of Balamand (CNRS/UOB Grant ref. 02-02-18). The authors would like to acknowledge the National Council for Scientific Research of Lebanon (CNRS-L), the Agence Universitaire de la Francophonie (AUF), and the Research Council at Saint Joseph University of Beirut (USJ, Project FS148) for granting a doctoral fellowship to Anna Maria Abi Khattar.

Acknowledgments: The authors would like to acknowledge Ziad Daoud for providing the bacterial strains and André El-Khoury for assistance with the antifungal analyses.

References

1. Ahmad-qasem, M.H.; Canovas, J.; Barrajon, E.; Carreres, J.E.; Micol, V.; Perez, J.V.G. Influence of olive leaf processing on the bioaccessibility of bioactive polyphenols. *J. Agric. Food Chem.* **2014**, *62*, 6190–6198. [CrossRef]
2. Lapage, S.P.; Sneath, P.H.A.; Lessel, E.F.; Skerman, V.B.D.; Seeliger, H.P.R.; Clark, W.A. *Foreword to the First Edition*; FAO: Rome, Italy, 2005.

3. Guinda, Á.; Castellano, J.M.; Delgado-hervás, T.; Gutiérrez-adánez, P.; Rada, M. Determination of major bioactive compounds from olive leaf. *LWT Food Sci. Technol.* **2015**, *64*, 431–438. [CrossRef]

4. Solarte-Toro, J.C.; Romero-Garcia, J.M.; Lopez-Linares, J.C.; Ramos, E.R.; Castro, E.; Alzate, C.A.C. Simulation Approach Through the Biorefinery Concept of the Antioxidants, Lignin and Ethanol Production using Olive Leaves as Raw Material. *Chem. Eng. Trans.* **2018**, *70*, 925–930.

5. Mkaouar, S.; Bahloul, N.; Gelicus, A.; Allaf, K.; Kechaou, N. Instant controlled pressure drop texturing for intensifying ethanol solvent extraction of olive (*Olea europaea*) leaf polyphenols. *Sep. Purif. Technol.* **2015**, *145*, 139–146. [CrossRef]

6. Khemakhem, I.; Gargouri, O.D.; Dhouib, A.; Ayadi, M.A.; Bouaziz, M. Oleuropein rich extract from olive leaves by combining microfiltration, ultrafiltration and nanofiltration. *Sep. Purif. Technol.* **2017**, *172*, 310–317. [CrossRef]

7. Romani, A.; Mulas, S.; Heimler, D. Polyphenols and secoiridoids in raw material (*Olea europaea* L. leaves) and commercial food supplements. *Eur. Food Res. Technol.* **2017**, *243*, 429–435. [CrossRef]

8. Dedousi, M.; Mamoudaki, V.; Grigorakis, S.; Makris, D.P. Ultrasound-Assisted Extraction of Polyphenolic Antioxidants from Olive (*Olea europaea*) Leaves Using a Novel Glycerol/Sodium-Potassium Tartrate Low-Transition Temperature Mixture (LTTM). *Environ. Artic.* **2017**, *4*, 31. [CrossRef]

9. Talhaoui, N.; Taamalli, A.; Mar, A. Phenolic compounds in olive leaves: Analytical determination, biotic and abiotic influence, and health benefit. *Food Res. Int.* **2015**, *77*, 92–108. [CrossRef]

10. Stamatopoulos, K.; Katsoyannos, E.; Chatzilazarou, A. Antioxidant Activity and Thermal Stability of Oleuropein and Related Phenolic Compounds of Olive Leaf Extract after Separation and Concentration by Salting-Out-Assisted Cloud Point Extraction. *Antioxidants* **2014**, *80*, 229–244. [CrossRef]

11. Mkaouar, S.; Gelicus, A.; Bahloul, N.; Allaf, K.; Kechaou, N. Kinetic study of polyphenols extraction from olive (*Olea europaea* L.) leaves using instant controlled pressure drop texturing. *Sep. Purif. Technol.* **2016**, *161*, 165–171. [CrossRef]

12. Galanakis, C.M.; Kotsiou, K. *Chapter 10—Recovery of Bioactive Compounds from Olive Mill Waste*; Elsevier: Amsterdam, The Netherlands, 2017.

13. Deng, J.; Xu, Z.; Xiang, C.; Liu, J.; Zhou, L.; Li, T.; Yang, Z.; Ding, C. Comparative evaluation of maceration and ultrasonic-assisted extraction of phenolic compounds from fresh olives. *Ultrason. Sonochem.* **2017**, *37*, 328–334. [CrossRef]

14. Rajha, H.N.; Mhanna, T.; El Kantar, S.; El Khoury, A.; Louka, N.; Maroun, R.G. Innovative process of polyphenol recovery from pomegranate peels by combining green deep eutectic solvents and a new infrared technology. *Lwt* **2019**, *111*, 138–146. [CrossRef]

15. Ahmad-Qasem, M.H.; Barrajón-Catalán, E.; Micol, V.; Mulet, A.; García-Pérez, J.V. In fl uence of freezing and dehydration of olive leaves (var. Serrana) on extract composition and antioxidant potential. *FRIN* **2013**, *50*, 189–196.

16. Japón-Luján, R.; Luque-Rodríguez, J.M.; de Castro, M.D.L. Multivariate optimisation of the microwave-assisted extraction of oleuropein and related biophenols from olive leaves. *Anal. Bioanal. Chem.* **2006**, *385*, 753–759. [CrossRef]

17. Şahin, S.; Samli, R.; Tan, A.S.B.; Barba, F.J.; Chemat, F.; Cravotto, G.; Lorenzo, J.M. Solvent-free microwave-assisted extraction of polyphenols from olive tree leaves: Antioxidant and antimicrobial properties. *Molecules* **2017**, *22*, 1056. [CrossRef]

18. Xynos, N.; Papaefstathiou, G.; Gikas, E.; Argyropoulou, A.; Aligiannis, N.; Skaltsounis, A. Design optimization study of the extraction of olive leaves performed with pressurized liquid extraction using response surface methodology. *Sep. Purif. Technol.* **2014**, *122*, 323–330. [CrossRef]

19. de Lucas, A.; De, E.M.; Rinco, J. Supercritical fluid extraction of tocopherol concentrates from olive tree leaves. *J. Supercrit. Fluids* **2002**, *22*, 221–228. [CrossRef]

20. Baldino, A.L.; della Porta, G.; Sesti, L.; Reverchon, E.; Adami, R. Concentrated Oleuropein Powder from Olive Leaves using Alcoholic Extraction. *J. Supercrit. Fluids* **2017**, *133*, 65–69. [CrossRef]

21. Escobedo, R.; Miranda, R.; Martínez, J. Infrared irradiation: Toward green chemistry, a review. *Int. J. Mol. Sci.* **2016**, *17*, 453. [CrossRef]

22. Rajha, H.N.; Debs, E.; Maroun, R.G.; Louka, N. Système d'extraction, de séparation ou de prétraitement assisté par rayonnement infrarouge. Adéquation entre les caractéristiques du rayonnement et celles de la matière traitée. Lebanese Patent 11296, 29 November 2017.

23. Raafat, K.; El-Darra, N.; Saleh, F.A.; Rajha, H.N.; Maroun, R.G.; Louka, N. Infrared-Assisted Extraction and HPLC-Analysis of Prunus armeniaca L. Pomace and Detoxified-Kernel and their Antidiabetic Effects. *Phytochem. Anal.* **2018**, *29*, 156–167. [CrossRef]

24. Cheaib, D.; El Darra, N.; Rajha, H.N.; El Ghazzawi, I.; Maroun, R.G.; Louka, N. Biological activity of apricot byproducts polyphenols using solid–liquid and infrared-assisted technology. *J. Food Biochem.* **2018**, *42*, 1–9.

25. Cheaib, D.; El Darra, N.; Rajha, H.N.; El-Ghazzawi, I.; Mouneimne, Y.; Jammoul, A.; Maroun, R.G.; Louka, N. Study of the Selectivity and Bioactivity of Polyphenols Using Infrared Assisted Extraction from Apricot Pomace Compared to Conventional Methods. *Antioxidants* **2018**, *7*, 174. [CrossRef]

26. Raafat, K.; El-Darra, N.; Saleh, F.A.; Rajha, H.N.; Louka, N. Optimization of infrared-assisted extraction of bioactive lactones from *Saussurea lappa* L. and their effects against gestational diabetes. *Pharmacogn. Mag.* **2019**, *15*, 208–218. [CrossRef]

27. Cifá, D.; Skrt, M.; Pittia, P.; di Mattia, C.; Ulrih, N.P. Enhanced yield of oleuropein from olive leaves using ultrasound-assisted extraction. *Food Sci. Nutr.* **2018**, *6*, 1128–1137. [CrossRef]

28. Singleton, V.L.; Lamuela-ravents, R.M. Analysis of Total Phenols and Other Oxidation Substrates and Antioxidants by Means of Folin-Ciocalteu Reagent. *METHODS Enzym.* **1999**, *299*, 152–178.

29. Mehta, D.; Sharma, N.; Bansal, V.; Sangwan, R.S.; Yadav, S.K. Impact of ultrasonication, ultraviolet and atmospheric cold plasma processing on quality parameters of tomato-based beverage in comparison with thermal processing. *Innov. Food Sci. Emerg. Technol.* **2019**, *52*, 343–349. [CrossRef]

30. Selin, C.; Zeynep, I. A novel approach for olive leaf extraction through ultrasound technology: Response surface methodology versus artificial neural networks. *Korean J. Chem. Eng.* **2014**, *31*, 1661–1667.

31. Iban, E.; Zarrouk, M. Optimization of Microwave-Assisted Extraction for the Characterization of Olive Leaf Phenolic Compounds by Using HPLC-ESI-TOF-MS/IT-MS 2. *J. Agric. Food Chem.* **2012**, *60*, 791–798.

32. Prieto, P.; Pineda, M.; Aguilar, M. Spectrophotometric Quantitation of Antioxidant Capacity through the Formation of a Phosphomolybdenum Complex: Specific Application to the Determination of Vitamin E 1. *Anal. Biochem.* **1999**, *341*, 337–341. [CrossRef]

33. Kallithraka, S.; Mohdaly, A.A. Determination of major anthocyanin pigments in Hellenic native grape varieties (*Vitis vinifera* sp.): Association with antiradical activity. *J. Food Compos. Anal.* **2005**, *18*, 375–386. [CrossRef]

34. Chanioti, S.; Tzia, C. Extraction of phenolic compounds from olive pomace by using natural deep eutectic solvents and innovative extraction techniques. *Innov. Food Sci. Emerg. Technol.* **2018**, *49*, 192–201. [CrossRef]

35. C.L.S. Institute. *Performance Standards for Antimicrobial Susceptibility Testing*; C.L.S. Institute: Wayne, PA, USA, 2018.

36. Stamatopoulos, K.; Chatzilazarou, A.; Katsoyannos, E. Optimization of Multistage Extraction of Olive Leaves for Recovery of Phenolic Compounds at Moderated Temperatures and Short Extraction Times. *Foods* **2014**, *3*, 66–81. [CrossRef]

37. Gómez-Caravaca, A.M.; Verardo, V.; Segura-Carretero, A.; Fernández-Gutiérrez, A.; Caboni, M.F. *Polyphenols in Plants: Isolation, Purification and Extract Preparation*; Elsevier: Amsterdam, The Netherlands, 2014.

38. Richardson, P. *Thermal Technologies in Food Processing*; Springer: Berlin, Germany, 2000.

39. Gorban, A.N.; Sargsyan, H.P.; Wahab, H.A. Quasichemical Models of Multicomponent Nonlinear Diffusion. *Math. Model. Nat. Phenom.* **2010**, *6*, 184–262. [CrossRef]

40. Silva, E.M.; Rogez, H.; Larondelle, Y. Optimization of extraction of phenolics from Inga edulis leaves using response surface methodology. *Sep. Purif. Technol.* **2007**, *55*, 381–387. [CrossRef]

41. Putnik, P.; Barba, F.J.; Španić, I.; Zorić, Z.; Dragović-Uzelac, V.; Kovačević, D.B. Green extraction approach for the recovery of polyphenols from Croatian olive leaves (*Olea europaea*). *Food Bioprod. Process.* **2017**, *106*, 19–28. [CrossRef]

42. Ahmad-Qasem, M.H.; Cánovas, J.; Barrajón-Catalán, E.; Micol, V.; Cárcel, J.A.; García-Pérez, J.V. Kinetic and compositional study of phenolic extraction from olive leaves (var. Serrana) by using power ultrasound. *Innov. Food Sci. Emerg. Technol.* **2013**, *17*, 120–129. [CrossRef]

43. Mylonaki, S.; Kiassos, E.; Makris, D.P. Optimisation of the extraction of olive (*Olea europaea*) leaf phenolics using water/ethanol-based solvent systems and response surface methodology. *Anal. Bioanal. Chem.* **2008**, *392*, 977–985. [CrossRef]

44. Gałuszka, A.; Migaszewski, Z.; Namieśnik, J. The 12 principles of green analytical chemistry and the SIGNIFICANCE mnemonic of green analytical practices. *TrAC Trends Anal. Chem.* **2013**, *50*, 78–84. [CrossRef]

45. Abaza, B.L.; Youssef, N.B.; Manai, H.; Haddada, F.M. Chétoui olive leaf extracts: Influence of the solvent type on phenolics and antioxidant activities. *Grasas Y Aceites* **2011**, *62*, 96–104. [CrossRef]
46. Sahin, S.; Samlı, R. Ultrasonics Sonochemistry Optimization of olive leaf extract obtained by ultrasound-assisted extraction with response surface methodology. *Ultrason. Sonochem. J.* **2013**, *20*, 595–602. [CrossRef]
47. Bilgin, M.; Selin, S. Effects of geographical origin and extraction methods on total phenolic yield of olive tree (*Olea europaea*) leaves. *J. Taiwan Inst. Chem. Eng.* **2013**, *44*, 8–12. [CrossRef]
48. Khemakhem, I.; HussamAhmad-Qasem, M.; Catalán, E.B.; Micol, V.; García-Pérez, J.V.; Ayadi, M.A.; Bouaziz, M. Kinetic improvement of olive leaves' bioactive compounds extraction by using power ultrasound in a wide temperature range. *Ultrason. Sonochem.* **2017**, *34*, 466–473. [CrossRef]
49. Bouallagui, Z.; Han, J.; Isoda, H.; Sayadi, S. Hydroxytyrosol rich extract from olive leaves modulates cell cycle progression in MCF-7 human breast cancer cells. *Food Chem. Toxicol.* **2011**, *49*, 179–184. [CrossRef]
50. Abaza, L.; Talorete, T.P.; Yamada, P.; Kurita, Y.; Zarrouk, M.; Isoda, H. Induction of Growth Inhibition and Differentiation of Human Leukemia HL-60 Cells by a Tunisian Gerboui Olive Leaf Extract. *Biosci. Biotechnol. Biochem.* **2007**, *71*, 1306–1312. [CrossRef]
51. Xie, P.; Huang, L.; Zhang, C.; Zhang, Y. Phenolic compositions, and antioxidant performance of olive leaf and fruit (*Olea europaea* L.) extracts and their structure—Activity relationships. *J. Funct. Foods* **2015**, *16*, 460–471. [CrossRef]
52. Korukluoglu, M.; Sahan, Y.; Yigit, A.; Ozer, E.T. Antibacterial activity and chemical constitutions. *Food Control* **2008**, *34*, 383–396.
53. El, S.N.; Karakaya, S. Olive tree (*Olea europaea*) leaves: Potential beneficial effects on human health. *Nutr. Rev.* **2009**, *67*, 632–638. [CrossRef]
54. Yuan, J.; Wang, C.; Ye, J.; Tao, R.; Zhang, Y. Enzymatic Hydrolysis of Oleauropein from *Olea europea* (Olive) Leaf Extract and Antioxidant Activities. *Molecules* **2015**, *11*, 2903–2921. [CrossRef]
55. Sahin, S.; Bilgin, M. Olive tree (*Olea europaea* L.) leaf as a waste by-product of table olive and olive oil industry: A review. *J. Sci. Food Agric.* **2018**, *98*, 1271–1279. [CrossRef]
56. El, A.; Rizk, T.; Lteif, R.; Azouri, H.; Delia, M.; Lebrihi, A. Fungal contamination and Aflatoxin B1 and Ochratoxin A in Lebanese wine—Grapes and musts. *Food Chem. Toxicol. J.* **2008**, *46*, 2244–2250.
57. Klan, A.; Jer, B.; Smole, S. Evaluation of diffusion and dilution methods to determine the antibacterial activity of plant extracts. *J. Microbiol. Methods* **2010**, *81*, 121–126.
58. Gökmen, M.; Kara, R.; Akkaya, L.; Torlak, E.; Önen, A. Evaluation of antimicrobial activity in olive (*Olea europaea*) leaf extract. *Am. J. Microbiol.* **2015**, *5*, 37–40.
59. Karygianni, L.; Cecere, M.; Skaltsounis, A.L.; Argyropoulou, A.; Hellwig, E.; Aligiannis, N.; Wittmer, A.; Al-Ahmad, A. High-Level Antimicrobial Efficacy of Representative Mediterranean Natural Plant Extracts against Oral Microorganisms. *BioMed Res. Int.* **2014**, *2014*, 839019. [CrossRef]
60. Daoud, Z.; Abdo, E.; Abdel-Massih, R.M. Antibacterial activity of rheum rhaponticum, olea europaea, and viola odorata on esbl producing clinical isolates of escherichia coli and klebsiella pneumoniae. *Int. J. Pharm. Sci. Res.* **2011**, *2*, 1669–1678.

 *antioxidants* 

Article

Apple Pomace Extract as a Sustainable Food Ingredient

Pedro A. R. Fernandes [1], Sónia S. Ferreira [1], Rita Bastos [1], Isabel Ferreira [2], Maria T. Cruz [2], António Pinto [3,4,5], Elisabete Coelho [1], Cláudia P. Passos [1], Manuel A. Coimbra [1], Susana M. Cardoso [1,*] and Dulcineia F. Wessel [1,3,4,*]

1   QOPNA & LAQV-REQUIMTE, Department of Chemistry, University of Aveiro, 3810-193 Aveiro, Portugal; pedroantonio@ua.pt (P.A.R.F.); soniasferreira@ua.pt (S.S.F.); ritabastos@ua.pt (R.B.); ecoelho@ua.pt (E.C.); cpassos@ua.pt (C.P.P.); mac@ua.pt (M.A.C.)
2   Faculty of Pharmacy and Center of Pharmaceutical Studies, University of Coimbra, Azinhaga de Santa Comba, 3000-548 Coimbra, Portugal; isabelcvf@gmail.com (I.F.); trosete@ff.uc.pt (M.T.C.)
3   School of Agriculture and CI&DETS, Polytechnic Institute of Viseu, 3500-606 Viseu, Portugal; apinto@esav.ipv.pt
4   CITAB, University of Trás-os-Montes e Alto Douro, 5001-801 Vila Real, Portugal
5   CERNAS, Agrarian Higher School, 3045-601 Coimbra, Portugal
*   Correspondence: susanacardoso@ua.pt (S.M.C.); ferdulcineia@esav.ipv.pt (D.F.W.); Tel.: +351-234-370-360 (S.M.C.); +351-232-480-703 (D.F.W.)

Received: 9 April 2019; Accepted: 16 June 2019; Published: 21 June 2019

Abstract: Apple pomace is a by-product of apple processing industries with low value and thus frequent disposal, although with valuable compounds. Acidified hot water extraction has been suggested as a clean, feasible, and easy approach for the recovery of polyphenols. This type of extraction allowed us to obtain 296 g of extract per kg of dry apple pomace, including 3.3 g of polyphenols and 281 g of carbohydrates. Ultrafiltration and solid-phase extraction using C18 cartridges of the hot water extract suggested that, in addition to the apple native polyphenols detected by ultra-high-pressure liquid chromatography coupled to a diode-array detector and mass spectrometry UHPLC-DAD-ESI-MS^n, polyphenols could also be present as complexes with carbohydrates. For the water-soluble polyphenols, antioxidant and anti-inflammatory effects were observed by inhibiting chemically generated hydroxyl radicals (OH•) and nitrogen monoxide radicals (NO•) produced in lipopolysaccharide-stimulated macrophages. The water-soluble polyphenols, when incorporated into yogurt formulations, were not affected by fermentation and improved the antioxidant properties of the final product. This in vitro research paves the way for agro-food industries to achieve more diversified and sustainable solutions towards their main by-products.

Keywords: polyphenols; antioxidant; anti-inflammatory; extraction; functional food

---

1. Introduction

The fact that agro-food industrial by-products are generally disposed, often with great expenses and environmental implications, has raised the need for their valuation [1]. Their perishable nature, due to the high-water content and huge amounts of organic load, as well as their chemical composition, particularly in dietary fiber and phytochemicals, provides a costless source of bioactive compounds that may favor an efficient and sustainable industrial development [1,2]. As a matter of fact, a circular economy model can be implemented in the agro-food sector by recycling its by-products, thereby creating added value with fewer resources.

Among the wide variety of agro-food industrial by-products available worldwide, apple pomace, resultant from apple (*Malus* spp., Rosaceae) processing, can be highlighted given the ubiquitous presence of the fruit in the diet of all cultures [3]. Actually, every year millions of tons of apples are processed to produce apple cider, juices, or concentrates, which yield huge amounts of residues, comprising the pulp, skin, seeds, and stalks from the fruit [2]. Several strategies for the valuation of apple pomace have been proposed, including its direct use for animal feed, organic acids, enzymes, bioethanol and biogas production by microbial fermentation, or the development of new materials as part of biocomposites [4]. Nevertheless, this by-product still presents a significant edible fraction which can be used as a source of valuable components. These may include hydroxycinnamic acids (chlorogenic acid and *p*-coumaroylquinic acid), flavan-3-ols (monomers such as epicatechin to large polymers known as procyanidins), flavonols (quercetin rutinoside, galactoside, glucoside, xyloside, arabinoside, and rhamnoside derivatives), dihydrochalcones (phloretin 2-*O*-glucoside and phloretin 2-*O*-xyloglucoside), and anthocyanins (cyanidin 3-*O*-galactoside) [2,5,6]. Furthermore, given the occurrence of polyphenol oxidation reactions, polyphenols might also be found as components attached to cell wall polysaccharides [7,8], which may have an impact on the antioxidant and antiviral properties attributed to apple pomace extracts [9]. Most studies aiming to evaluate the potential applications of apple pomace valuable components have been performed with the use of organic solvents, which may be appropriate for pharmaceutical or cosmetic purposes but not for food applications.

Therefore, this work aims to give new insights into the nature of the apple pomace water-soluble polyphenols and their bioactivity, as well as to evaluate the potential of its water extract to be used for the supplementation or development of fortified products. To this end, in addition to apple native polyphenols, the occurrence of polyphenol/carbohydrate complexes was inferred employing ultrafiltration and solid-phase extraction of the hot water extract. Furthermore, the polyphenols isolated by solid-phase extraction were also used to provide evidence of their antioxidant and anti-inflammatory properties using both chemical and cellular inflammatory models. The feasibility of incorporating the aqueous extract of apple pomace into foods was tested by its addition to yogurt formulations and its potential impact on the fermentation process (pH, titratable acidity, and lactic acid counting), and antioxidant and nutritional properties of the final product.

2. Materials and Methods

2.1. Chemicals

All reagents were of analytical grade. All standard compounds used for polyphenol quantification by UHPLC-DAD-ESI-MS^n or antioxidant assays had a purity level of at least 95%.

2.2. Preparation of Extracts

Agro-industrial apple pomace was obtained following the general procedure described by Kennedy et al. [10]. This process consisted of the processing of a mixture of apples, mainly composed of the Royal Gala variety, employing milling, enzymatic digestion (amylase, pectin lyase, and polygalacturonase), and pressing processes for a period of at least 3 hours. After processing, the apple pomace was frozen at −20 °C, freeze-dried, sealed in bags, and stored in a dark at room temperature in a desiccator until further analysis. Extracts were prepared from apple pomace, using boiling water with 1% acetic acid, pH 2.5, at a solid (dry weight) to a solvent ratio of 1:60 (g/mL), the optimal conditions for polyphenol extraction as determined by Çam and Aaby [5]. The procedure was limited to a period of 10 min as the extraction yields (in mass) hardly improved for more than 10% using longer periods, e.g., 1 or 2 h, and allowed us to avoid the polyphenols thermal degradation [11,12]. Afterwards, the extracts were filtered (Whatman filter paper n° 4 and G3 sintered funnel), and the residue was re-extracted two more times following the same procedure to recover any remnant material. The crude extracts were combined, concentrated under reduced pressure, and freeze-dried, yielding a hot water extract (HWE). For characterization of the high and low molecular weight material (Figure 1), the HWE was fractionated at room temperature on an ultrafiltration module – Labscale TFF System (Merck KGaA,

Darmstadt, Germany), using a pellicon XL ultrafiltration ultracel membrane with cut-off 10 kDa, as previously described by Passos et al. [13].

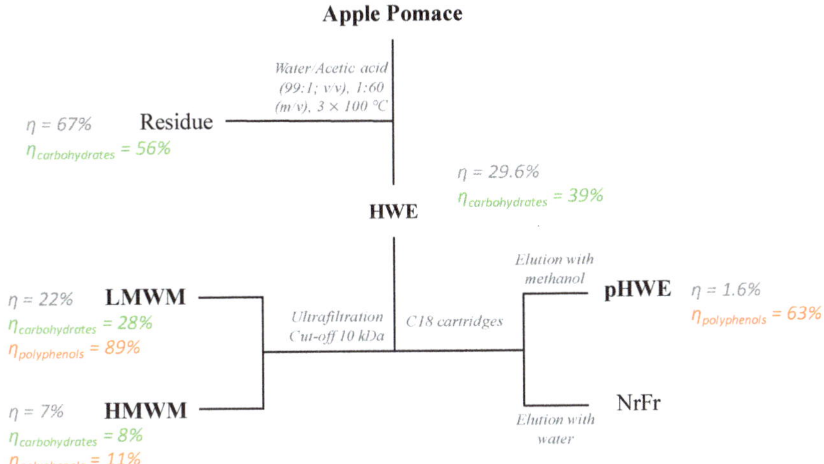

Figure 1. Schematic representation of the fractionation processes adopted in this work and yields of the extracts and polysaccharides in relation to apple pomace and polyphenols present in the hot water extract (dry basis). In bold are highlighted the fractions that were further studied. HWE—hot water extract; LMWM—low molecular weight material; HMWM—high molecular weight material; pHWE—purified hot water extract; NrFr—non-retained fraction.

To further characterize the polyphenolic composition and potential bioactive effects, the HWE was submitted to solid-phase extraction (Figure 1) in Sep-Pak C18 cartridges (SPE-C18, Supelco-Discovery (St. Louis, MO, USA, 20 g). The column was preconditioned with 20 mL of methanol followed by 20 mL of water. Afterwards, the sample was loaded onto the column, and the non-retained material (NrFr) was eluted with water, three times the volume of the cartridge. The material retained was eluted using methanol following the same procedure. The resultant polyphenol-isolated HWE fraction (pHWE) was concentrated under reduced pressure to remove the methanol and then frozen and freeze-dried.

For the fractionation processes by ultrafiltration or solid-phase extraction using C18 cartridges, polyphenols yields were estimated by mass balance between those initially found in the hot water extract and those obtained in the further sub-fractions. For ultrafiltration it was estimated as described in Equation (1):

$$PCHWE = PCHMWM + PCLMWM \tag{1}$$

where, PCHWE, PCHMWM, and PCLMWM corresponds to the mass of polyphenols present in the HWE, high molecular weight material fraction, and low molecular weight material fractions, respectively.

For solid phase extraction the yields were estimated considering Equation (2):

$$PCHWE = PCpHWE + PCNrFr \tag{2}$$

where PCHWE, PCpHWE, and PCNrFr corresponds to the mass of polyphenols present in the HWE, in the fraction retained in the C18 cartridges (pHWE), and in the non-retained fraction in the C18 cartridge (NrFr), respectively. For polysaccharides, a similar rationale was applied taking as a reference the number of polysaccharides initially present in the apple pomace.

2.3. General Chemical Characterization

The moisture content of apple pomace was determined by the weight difference before and after drying for 12 h at 105 °C, up to constant weight. For apple pomace and derived extracts, protein was estimated by determining total nitrogen using a Truspec 630-200-200 elemental analyzer (St. Joseph, MI, USA) with a thermal conductivity detector (TDC) and employing a conversion factor of 5.72, as estimated for apples [14]. To obtain quantitative and qualitative information, the carbohydrate composition of the samples was determined by adopting the general procedure of Fernandes et al. [8]. Briefly, neutral sugars were determined using gas chromatography (GC) analysis after acid hydrolysis (12 M H_2SO_4 for 3 h at room temperature, followed by 2.5 h hydrolysis in 1 M H_2SO_4 at 100 °C), reduction with $NaBH_4$ (15% w/v in 3 M NH_3 during 1 h at 30 °C), and acetylation (with acetic anhydride in the presence of 1-methylimidazole during 30 min at 30 °C). Uronic acids were quantified by the 3-phenylphenol colorimetric method after acid hydrolysis (1 h in 1 M H_2SO_4 at 100 °C) of the sample [15]. Galacturonic acid (GalA) was used as standard. Free sugars were determined following the same procedure without the hydrolysis step. Fructose was quantified as the sum of mannitol and sorbitol due to the epimerization of fructose during the reduction step, using the ratio of the epimerization reaction [16]. The amount of total polyphenolic compounds was quantified using the Folin–Ciocalteu method [17] using concentrations from 1 to 10 mg of extract/mL water. The results were expressed as g gallic acid equivalent (GAE)/kg of extract.

Individual polyphenols were determined by UHPLC-DAD-ESI-MS^n on an Ultimate 3000 (Dionex Co., San Jose, CA, USA) apparatus equipped with a Diode Array Detector (Dionex Co., USA) and coupled to a mass spectrometer. The chromatographic system consisted of a quaternary pump, an autosampler, a degasser, a photodiode-array detector, and an automatic thermostatic column compartment. Hypersil Gold (Thermo Scientific, San Jose, CA, USA) C18 column (100 mm length; 2.1 mm i.d.; 1.9 μm particle diameter, end-capped) at 30 °C was used. The mobile phase was composed of (A) 0.1% (v/v) formic acid and (B) acetonitrile. The solvent gradient started with 5% of solvent (B), reaching 40% at 14 min and 100% at 16 min, followed by the return to the initial conditions. The flow rate was 0.1 mL min^{-1} and UV-Vis spectral data for all peaks were accumulated in the range 200–500 nm, while the chromatographic profiles were recorded at 280, 320, and 340 nm for polyphenol analysis. The mass spectrometer used was a Thermo LTQ XL (Thermo Scientific, USA) ion trap MS, equipped with an electrospray ionization (ESI) source. Control and data acquisition were carried out with the Thermo Xcalibur Qual Browser data system (Thermo Scientific, USA). Nitrogen above 99% purity was used, and the gas pressure was 520 kPa (75 psi). The instrument was operated in negative-ion mode with ESI needle voltage set at 5.00 kV and an ESI capillary temperature of 275 °C. The full scan covered the mass range from m/z 100 to 2000. CID–MS/MS and MS^n experiments were acquired for precursor ions using helium as the collision gas with an energy of 25–35 arbitrary units.

For quantitative analysis, calibration curves were performed by injection of 5 known concentrations of standard compounds. Detection (LOD) and quantification (LOQ) limits were calculated using the parameters of the calibration curves, defined as 3.3 and 10 times the value of the regression error divided by the slope, respectively.

For rutin (ACROS), the tested range was 1.0–10.0 μg/mL, and the equation was y = 12,624x − 953, with an R^2 value of 0.999. The quantification limit (LQ) and detection limit (LD) for this compound were 1.29 and 0.43 μg/mL, respectively. For quercetin-3-*O*-glucoside (Sigma-Aldrich), the tested range was 2.4–12.2 μg/mL, the equation was y = 16,421x − 879 with an R^2 value of 0.999. The LQ and LD were 1.07 and 0.35 μg/mL, respectively. The calibration curve of phloridzin (Sigma-Aldrich), y = 20,429x − 456, were performed for ranges of 2.3–11.7 μg/mL, presenting an R^2 of 0.999. The determined LQ and LD were 1.35 and 0.45 μg/mL, respectively. The remaining quercetin derivatives were quantified as quercetin-3-*O*-glucoside equivalents.

2.4. Formulation of a Hot Water Extract (HWE)-Fortified Yogurt

The apple pomace HWE was used as an ingredient for yogurt formulations. Yogurts were prepared from ultra-high temperature pasteurized milk (composed of 5.1% of carbohydrates, 3.4% of protein, and 1.6% of fat) and milk powder at 1% w/w of milk (composed of 54% of carbohydrates, 34.5% of protein, and 1% of fat) in the absence of, or alternatively with the addition of, extract, to yield a control yogurt and a supplemented yogurt, respectively. A ratio of 3.3% (w of extract/w of milk) was used based on the maximum amount of extract soluble in milk after heating to 90 °C for 2 min and leaving to cool to 40 °C. Plain yogurt (composed of 4.0% of carbohydrates, 3.2% of protein, and 2.9% of fat), purchased at the local market, was added (1% w/w of milk) as inoculum to achieve a final Lactic Acid Bacteria count of 6 Log colony forming units (CFU)/g of mixture. This amount was determined based on the *Streptococcus thermophilus* counts present in the commercial yogurt and detected in M17 (Liofilchem, Rosetodegli Abruzzi, Italy), a medium specific for the growth of this bacterium. An incubation period of 72 h at 37 °C was used for *Streptococcus thermophilus* counting. For yogurt production, the mixture was incubated at 42 °C until reaching a pH below 4.5. The fermentation process was controlled by measuring the pH, titratable acidity (g of lactic acid/100 g), and *Streptococcus thermophilus* counts every 2 hours. For titratable acidity, samples were homogenized in water at a proportion of 1:9 (w/v). Afterwards, the pH value of the samples was measured, and titrated with 0.1 M NaOH in the presence of a few drops of phenolphthalein (1%) as an indicator. The titratable acidity was expressed in g of lactic equivalents/100 g of yogurt.

2.5. Antioxidant Activity

HWE and pHWE antioxidant activity was screened by the 2,2-diphenyl-1-picrylhydrazyl (DPPH•) [18] and 2,2′-azinobis-(3-ethylbenzothiazoline-6-sulfonic acid) (ABTS•$^+$) methods [19], using water as a solvent. The results were expressed as half maximum effective concentration (EC_{50}) (µg/mL), which represents the amount of extract required to reduce the radical concentration to half of its initial concentration. In addition, pHWE was evaluated for its capability to inhibit OH• radicals generated by the ferric-ascorbate-EDTA-H_2O_2 Fenton system, following the general procedure of Kunchandy and Rao [20]. The scavenging of OH• was measured by determining the relative amount of oxidized deoxyribose formed in the presence and absence of the extract. The results were expressed as mannitol equivalents (mmol/g of extract). As the OH• scavenging is based on the inhibition of deoxyribose oxidation by antioxidants, other sugars present, such as those found in the HWE, could interfere. For this reason, the antioxidant activity was measured on pHWE, obtained by purification of the HWE by solid-phase extraction.

To evaluate the effect of HWE addition to yogurts, total polyphenolic content and antioxidant activity, as well as their stability along the fermentation process, the control and supplemented yogurt with the HWE were individually extracted twice with methanol/water/acetic acid solutions (80:19:1; v/v/v) along different fermentation times. The resulting extracts were concentrated, freeze-dried, and analyzed by the Folin–Ciocalteu protocol (µg GAE/100 g of yogurt fresh weight). For the antioxidant activity, the ABTS•$^+$ method (µg Trolox equivalents/100 g of yogurt fresh weight) was selected given its simplicity. The same concentrations and solvents as those described for the HWE and pHWE extracts were used.

2.6. Anti-Inflammatory Potential

2.6.1. Inhibition of Chemically-Induced NO• Production

The NO• scavenging method was adapted from Catarino et al. [21]. Briefly, 100 µL of the pHWE, solubilized in phosphate buffer at pH 7.4 at (260 µg/mL), was mixed with 100 µL of sodium nitroprusside 3.33 mM (also in buffer) and incubated under a fluorescent lamp (Tryun 26 W) for 15 min. Afterwards, 100 µL of Griess reagent (0.5% sulphanilamide and 0.05% naphthyletylenediamine dihydrochloride in 2.5% H_3PO_4) was added, and the mixture was incubated in the dark for 10 min. The absorbance was measured at 562 nm. NO• scavenging was expressed as % of inhibition.

2.6.2. Inhibition of NO• Production in LPS-Stimulated Macrophages Cell Line Raw 264.7

The extracts were solubilized in sterile phosphate-buffered saline (PBS) with 2% (v/v) dimethyl sulfoxide (DMSO) and filtered through a cellulose acetate 0.22 μm sterile syringe filter (Frilabo, Maia, Portugal) under sterile conditions. The solutions were then diluted to achieve 281–1490 μg/mL of pHWE in the culture medium, with a final concentration of dimethyl sulfoxide (DMSO) lower than 0.1% (v/v). The medium was composed of Dulbecco's Modified Eagle Medium (DMEM, A13169050, Applichem, Darmstadt, Germany) supplemented with 10% non-inactivated fetal bovine serum (Alfagene, Carcavelos, Portugal), 100 U/mL penicillin, 100 μg/mL streptomycin, and 17.95 mM sodium bicarbonate, all from Sigma, St. Louis, MO, USA.

Raw 264.7 cells, a mouse leukemic monocyte macrophage cell line from American Type Culture Collection (ATCC TIB-71), were plated (3×10^5 cells/well) and allowed to stabilize for 12 h. Afterwards, the cell medium was replaced, and the cells were pre-incubated with 50 μL of pHWE or phosphate buffer with or without (control) 0.1% DMSO, for 1 h. Raw 264.7 cells were later activated with 1 μg/mL lipopolysaccharide (LPS from Escherichia coli, serotype 026:B6, Sigma Chemical Co., St. Louis, MO, USA) for 24 h. Cell viability was assessed using 3-(4,5-Dimethylthiazol-2-yl)-2,5-diphenyl tetrazolium bromide (MTT, Acros Organics, Geel, Belgium). The NO• production was determined by a colorimetric reaction with the Griess reagent, as previously reported by Búfalo et al. [22].

2.7. Nutritional Properties of the Yogurt

Given that apple pomace water extracts display high carbohydrate contents [23], the nutritional properties of the control and the HWE yogurts were evaluated by measuring the total sugar content and the amount of reducing sugars using the phenol-sulfuric method [24] and the 3,5-dinitrosalicylic acid (DNS) method [25]. The results were expressed as lactose equivalents/100 g, as lactose is the main carbohydrate found in dairy products. To complete the data set, moisture and protein contents were determined, as previously described. The protein conversion factor (6.15) estimated for dairy products was used for protein quantification [14]. The ash content was assessed by determining the final residue after incineration at 500 °C for 3 h. Fat was calculated by difference. The energetic value was calculated (Equation (3)) according to the energetic parameter published by the European Parliament [26]. As water extracts are known to present polysaccharides, their energetic contribution (2 kcal/g) was also included, assuming that polysaccharides are not affected and are not consumed by lactic acid bacteria during fermentation:

$$\text{Energy (kcal)} = 4 \times (\text{g of protein} + \text{g of reducing sugars}) + 9 \times (\text{g of lipids}) \\ + 2 \times (\text{g of added apple pomace aqueous extract polysaccharides}) \tag{3}$$

2.8. Statistical Analysis

All experiments were performed with at least three independent assays being represented as mean ± standard error of the mean. Data were statistically analyzed by a trial version of GraphPad Prism 6.01 software (OriginLab Corporation, Northampton, MA, USA) by one-way analysis of variance (ANOVA) followed by Tukey's multiple comparison test. A factor of 0.001 was used, unless otherwise stated.

3. Results

3.1. Apple Pomace Extracts

The industrial apple pomace had a high water-content (81%), rendering high perishability to this by-product. Protein comprised 50 g/kg of its dry weight, while carbohydrates were the major components (720 g/kg dry weight basis) (Table 1). These included the 180 g/kg of free sugars, mainly Fru (77 mol %), and 530 g/kg of polysaccharides with Glc (41 mol %), GalA (19 mol %), Ara (12 mol %), Xyl (10 mol %), and Gal (9 mol %) being the main sugars. This composition was characteristic of apple polysaccharides and reflects the presence of pectic polysaccharides as soluble dietary fiber and hemicelluloses and cellulose as insoluble dietary fiber [27,28]. Alongside these carbohydrates, polyphenols were also present. Therefore, hot water extraction was performed as it represents a cheap, non-toxic, environmentally friendly extraction procedure and is easily implementable at an industrial scale [5], in contrast to extractions using common organic solvents. To prevent polyphenol oxidation, diluted acetic acid was used [29].

HWE represented 29% of dry apple pomace and presented 11 g/kg of polyphenols (Table 1). This resulted from the co-extraction of carbohydrates (950 g/kg) and protein (13 g/kg), which accounted for 39% and 7% of that initially present in the apple pomace (Figure 1 and Table 1). The mass balance indicated that about 3.26 g GAE of polyphenols per kg of apple pomace were extracted. These yields were higher than those obtained with water at room temperature (2.6 g/kg of apple pomace) [30], and lower when using methanol (3.6 g/kg of apple pomace) or acetone (6.48 g GAE/kg of apple pomace) [9]. It is known that polysaccharides may interact with polyphenols, impairing their transfer from the fruit to the water fraction [31]. However, when using methanol and acetone, solvents that are of chaotropic nature, these interactions are disrupted and polyphenols become soluble [32], explaining the higher yields for methanol and acetone.

In order to infer about possible polyphenols–polysaccharides interactions in the HWE, ultrafiltration was performed using a 10 kDa ultrafiltration membrane. This process is based on the principle that apple polyphenols present, on average, a degree of polymerization of 5 [33], they would diffuse along the membrane unless being retained by any interaction phenomenon. According to the data presented in Table 1, 11% of the polyphenols from the HWE remained in the high molecular weight fraction. This fraction accounted for 6.9% of the apple pomace and was highly rich in polysaccharides (777 g/kg), mostly composed of Ara (35 mol %), GalA (36 mol %), and Glc (9 mol %). Ara and GalA are carbohydrates characteristic of pectic polysaccharides [28] while Glc is generally associated to glucans [23]. Therefore, it is possible to infer that the polyphenols present in the high molecular weight fraction were retained as a result of interactions with pectic polysaccharides and glucan fractions, as reported to occur between wine polyphenols and polysaccharides [34,35]. However, it is feasible that this retention mostly occurred as a result of covalent interactions between polyphenols and polysaccharides, due to the reactions between polyphenol quinones, formed by oxidation reactions and nucleophilic compounds of the cell wall [29]. In such case, polyphenols may establish bridges between different polysaccharide structures yielding chimeric structures [7]. Therefore, it is also possible to infer that some of the polyphenols present in the low molecular weight material (75% w/w of the HWE) were probably covalently associated to carbohydrates. Sugar analysis of this fraction showed the prevalence of GalA (45 mol %) and Ara (15 mol %), thus suggesting that they were associated to pectic oligosaccharides that globally represented 250 g/kg of the low molecular weight material. Such complexes are also present in the final extraction residue where polysaccharides represented 595 g/kg. Given the detection of GalA (13 mol%), Ara (8 mol%) and Gal (8 mol%), Glc (51 mol%), and Xyl (14 mol%), it is feasible that in addition to pectic polysaccharides, polyphenols were covalently linked to glucans [28,36].

Table 1. Yield (%), monosaccharide (molar%), carbohydrate (%), protein (%), and polyphenolic composition (g gallic acid equivalents (GAE)/kg) of industrial apple pomace, hot water extract (HWE), high molecular weight material (HMWM), low molecular weight material (LMWM), and extraction residue.

Fractions	Yield (%)	Yield of Carbohydrate (%)		Carbohydrate (mol%)									Total Carbohydrate (%)	Total Protein (%)	Total PC (g GAE/kg)
				Rha	Fuc	Ara	Xyl	Man	Gal	Glc	Fru	GalA			
Apple pomace			Polysaccharides	1	1	12	10	5	9	41		19	53.4	71.7	5.2	ND
			Free Sugars							23	77		18.3			
HWE	29.6	39.2	Polysaccharides	3	1	25	3		10	9		50	42.9	94.9	1.3	11
			Free Sugars							18	82		52.0			
HMWM	6.9	7.5	Polysaccharides	1	t	35	6	1	10	9		36	77.7	77.7	ND	5
LMWM	22.3	27.7	Polysaccharides	1	t	15	2		6	34		42	33.5	89.0	ND	9
			Free Sugars							6	94		55.5			
Residue	67.4	55.9	Polysaccharides	1	1	8	14	4	8	51		13	59.5	59.5	7.1	

t = trace; ND = not determined.

To better understand the nature of the polyphenols and their bioactivity, the HWE was subjected to solid-phase extraction based on the principle that polyphenols, which have hydrophobic character, would be retained on the C18 cartridge while carbohydrates, which are hydrophilic molecules, would elute from the cartridge. The hydrophobic fraction, named pHWE, corresponded to 1.6% of the dry apple pomace. According to Folin–Ciocalteu's method, polyphenols represent 149 g/kg and accounted for 63% of those in the HWE. The remaining 37% of the polyphenols eluted from the cartridge alongside with the HWE carbohydrates. These probably have the contribution of free sugars, which are known to have a reducing capacity and therefore interfere in the Folin–Ciocalteu method. However, the hypothesis that polyphenolic structures exist in this fraction as a result of covalent bonding to polysaccharides should not be discarded.

UHPLC-DAD-ESI-MS^n analysis showed that the pHWE was mainly composed of flavonols (Table 2) which included quercetin-3-*O*-galactoside (27%), quercetin-3-*O*-rhamnoside (23%), quercetin 3-*O*-arabinofuranoside (13%), and the dihydrochalcone phloretin-2-*O*-glucoside (14%). Quercetin-3-*O*-xylanoside (8%), quercetin-3-*O*-glucoside (7%), quercetin-*O*-pentoside (3%), quercetin-3-*O*-rutinoside (3%), and quercetin 3-*O*-arabinopyranoside (2%) were also detected as minor compounds, which was in agreement with previously reported work concerning the polyphenolic composition of apple pomace [5,37]. However, this analysis only explained 77% (115 mg/g of extract) of the polyphenols detected by the Folin–Ciocalteu method (149 mg/g of extract). According to Millet et al. [38], this difference, together with the fact that only 11% of the extract composition was explained, is highly suggestive of the occurrence of polyphenol oxidation products formed during apple processing [39]. Oxidation products of polyphenols are formed in very low amounts and present newly formed linkages [40], hardly quantitative by common ultra-high-pressure liquid chromatography (UHPLC) techniques [41]. These include hydroxycinnamic acid, dihydrochlacones, and flavan-3-ols oxidations products, already shown to be present in apple pomace [8].

Table 2. Retention time (RT), mass spectrum (MS), and polyphenolic composition (mg/g of extract) of pHWE.

N°	RT	Compound	λ_{max}	MS (*m/z*)	MS^2 (*m/z*)	Extract pHWE
1	12.3	Quercetin-3-*O*-rutinoside [a]	254, 353	609	463, 301	3.27 ± 0.06
2	12.6	Quercetin-3-*O*-galactoside [b]	256, 354	463	301	31.37 ± 0.32
3	12.7	Quercetin-3-*O*-glucoside [a]	256, 353	463	301	8.45 ± 0.10
4	13.2	Quercetin-3-*O*-xylanoside [b]	256, 354	433	301	8.88 ± 0.10
5	13.4	Quercetin 3-*O*-arabinopyranoside [b]	243, 352	433	301	2.31 ± 0.04
6	13.5	Quercetin 3-*O*-arabinofuranoside [b]	256, 352	433	301	15.09 ± 0.16
7	13.7	Quercetin-*O*-pentoside [b]	256, 351	433	301	3.32 ± 0.05
8	13.8	Quercetin-3-*O*-rhamnoside [b]	256, 350	447	301	26.05 ± 0.27
9	14.9	Phloretin-2-*O*-glucoside [a]	227, 284	435	273	15.96 ± 0.20
					Total	114.75 ± 1.25

Identification was performed based on (a) the corresponding standard; (b) UV and MS^n spectra, plus elution order previously described in the literature [5,37].

3.2. Antioxidant and Anti-Inflammatory Potential

To provide evidence of the antioxidant potential of aqueous extracts from apple pomace, two widespread chemical models, the DPPH• and ABTS•$^+$ radical inhibition assays, were used. As represented in Table 3, the EC$_{50}$ values of the HWE for the DPPH• and ABTS•$^+$ methods were 1.34 and 0.53 mg/mL, respectively. The increment of polyphenols in the purified fraction (pHWE) was reflected on the extract's antioxidant activity, which exhibited DPPH• and ABTS•$^+$ EC$_{50}$ of 82.4 and 35.2 µg/mL, respectively (Table 3). When expressing the antioxidant activity with reference to total polyphenols (Table 3), it was observed that pHWE accounted for 92% of the HWE antioxidant activity. These results suggested that most of the compounds responsible for the antioxidant properties of the HWE were recovered after purification using C18 cartridges. The pHWE also presented the capability to inhibit the formation of OH$\bullet$ (6.75 mannitol equivalents/g of extract) generated by Fenton reactions from a ferric-ascorbate-EDTA-H_2O_2 system [20]. This can be attributed to both iron chelation and direct scavenging of OH$\bullet$ by polyphenols. However, in contrast to DPPH$\bullet$ and ABTS$\bullet^+$ which are not biological radicals, OH$\bullet$ is present in biological systems, resulting from Fenton reactions and cellular processes such as cell respiration and inflammation, prone to damage cellular lipids, proteins, and nucleic acids [42]. Closer extrapolations to in vivo effects could be inferred when complementing with other radical generating systems, either using enzymes such as xanthine/xanthine oxidase or cellular models such as activated neutrophils [43].

Nitric oxide (NO$\bullet$) has been recognized as a versatile player in several biological mechanisms, including endothelial cell function and inflammation, turning it into a biomarker in the screening of new anti-hypertensive and anti-inflammatory drugs [44]. Therefore, the potential capability of the aqueous extracts from apple pomace to regulate NO$\bullet$-driven processes was inferred through its ability to scavenge chemically generated NO$\bullet$. As represented in Table 3, at a concentration of 130 µg/mL, the pHWE inhibited 35% of the NO$\bullet$ chemically generated, which is in accordance with the antioxidant properties previously described in this work. This inhibition was, however, lower than that produced by ascorbic acid (57%).

Table 3. Total polyphenolic content (TPC) and antioxidant (DPPH$\bullet$, ABTS$\bullet^+$, OH$\bullet$, NO$\bullet$) activity of the hot water extract before (HWE) and after the purification (pHWE).

Extract	TPC	DPPH$\bullet$	ABTS$\bullet^+$	OH$\bullet$	NO$\bullet$
HWE	10.7 ± 0.2	1339 ± 211 (14.2 ± 1.7)	532 ± 11.5 (5.69 ± 0.12)	-	-
pHWE	149 ± 1.87	82.4 ± 11.2 (12.3 ± 1.7)	35.2 ± 5.9 (5.23 ± 0.44)	6.75 ± 0.45	35.2 ± 5.9
AA	-	2.70 ± 0.30	2.68 ± 0.03	-	57.3 ± 2.3

The first and second values for the DPPH$\bullet$ and ABTS$\bullet^+$ are expressed in terms of EC_{50} (µg of extract/mL) and as relative antioxidant capacity with reference to total polyphenols (µg GAE of extract/mL), respectively. OH$\bullet$ scavenging was expressed as mannitol equivalents (mmol/g of extract), and for the NO$\bullet$ method as percentage of inhibition at 130 µg/mL. Values are compared to ascorbic acid (AA).

To provide closer evidence of anti-inflammatory effects occurring in vivo, which could potentiate the valuation of apple pomace for the development of functional foods, the pHWE was tested on LPS-stimulated Raw 264.7 cells, by measuring their effect on the accumulation of nitrites in the culture medium, an indicator of NO$^\bullet$ production. Indeed, macrophages activated by the Toll-like receptor 4 (TLR4) agonist LPS, produce large amounts of NO$\bullet$ and constitute a well-described in vitro model of inflammation, useful for the screening of molecules with anti-inflammatory activity [44]. In a first approach, the occurrence of possible cytotoxic effects triggered by the pHWE was evaluated by determining the cellular viability of Raw 264.7 macrophages, stimulated with LPS (Figure 2a). The presence of 0.1% DMSO did not affect the cell viability which was similar ($p > 0.05$) to the control in all tested concentrations. Accordingly, previously reported data showed that pure quercetin-glycosylated derivatives and phloridzin are not cytotoxic for the corresponding concentrations herein tested, as evaluated in similar cell models [45,46].

The anti-inflammatory potential of the pHWE was measured by the reduction of nitrite accumulated in the culture medium in comparison with the amount released by untreated LPS-stimulated Raw 264.7 cells. When treated with LPS, NO$\bullet$ released into the culture medium by macrophages increased from the basal value of 0.2 µM to 22 µM. Yet, when pre-incubated with pHWE in all non-cytotoxic concentrations (280, 375, 745, and 1490 µg/mL), the NO$\bullet$ released by macrophages was limited to 26%–84% of the value observed for non-treated cells (Figure 2b) in a concentration-dependent manner.

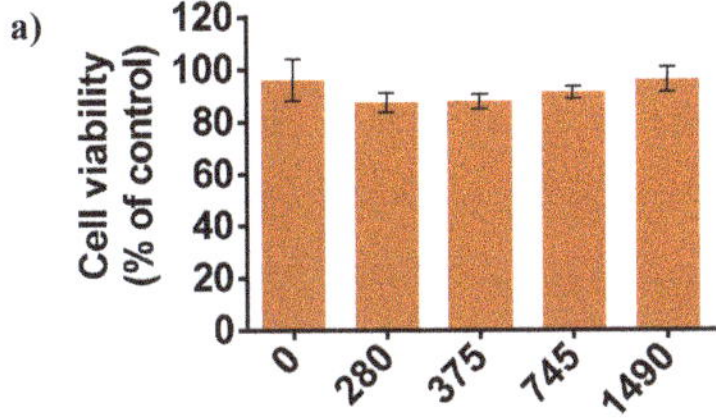

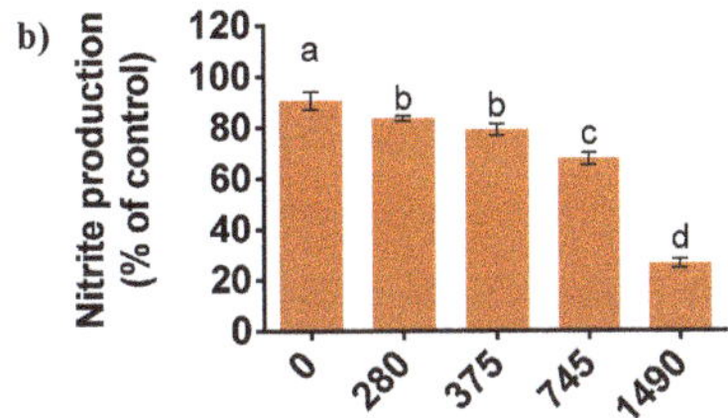

Figure 2. Treatment of mouse macrophage cell line, Raw 264.7, with apple pomace extracts, followed by incubation with lipopolysaccharide (LPS) from *Escherichia coli* as in vitro model of inflammation. (**a**) Cell viability (% of the control) of Raw 264.7 cells after incubation with polyphenolic-rich hot water extract (pHWE) at 0–1490 μg/mL in phosphate buffer/dimethyl sulfoxide (DMSO) (99.9:0.01; v/v). (**b**) Inhibitory effect of pHWE on LPS-induced nitrite production (% of the control) in Raw 264.7 cells. Data represent mean ± standard deviation of 3 independent assays. Different letters indicate statistical significance between pHWE concentrations (a,b,c, and d, $p < 0.001$) compared to control by one-way ANOVA followed by Tukey's Multiple comparison test.

Some of the major polyphenols present in the apple pomace aqueous extracts have been previously reported to possess anti-inflammatory properties. For example, quercetin, a flavonol, and some of its glycosylated derivatives were shown to inhibit NO• production in LPS-induced Raw 264.7 cells and to modulate several inflammatory signaling cascades [31,33]. Phloridzin, a dihydrochalchone, has also been described to modulate inflammatory responses [32]. Nevertheless, direct relation of the observed anti-inflammatory effect with the presence of polyphenols is still not possible to establish, since despite pHWE purification, other unknown compounds representing 85% of the extract, possibly apple polyphenol oxidation products formed during processing [29], may also be responsible for the effects. Furthermore, no comparisons with the HWE could be established given its poor solubility in phosphate-buffered saline with 2% (v/v) dimethyl sulfoxide (DMSO) at room temperature and the adverse effects on the viability of Raw 264.7 cells when using higher concentrations of DMSO. Although a deeper consolidation through the measurement of pro- and anti-inflammatory interleukins and/or other anti-inflammatory markers is required, these results evidence the possible valuation of apple pomace as a potential anti-inflammatory nutraceutical.

3.3. Application as a Food Ingredient in Yogurt Formulation

Given the presence of polysaccharide/polyphenol complexes and the possible antioxidant and anti-inflammatory properties that could be attributed to apple pomace extracts, their potential to be used as ingredients for food formulations was tested. Yoghurt was chosen as it is a worldwide, ready-to-eat product with a high nutritional value and positive bioactive effects that can be reinforced by the addition of other components [47–49]. As a result, HWE was incorporated into yogurt formulations to reach 3.3% ($w_{extract}/w_{milk}$). The incorporation of the HWE resulted in a mixture with an initial pH of 5.34, 19% lower than the control (6.56), which is in accordance with the higher titratable acidity observed (0.361 versus 0.097 g lactic acid/100 g of yogurt) (Figure 3a,b). This effect was similarly observed when adding wine grape pomace to yogurts [47] and could be related to naturally occurring organic acids, such as malic acid, in apple pomace. The similar number of *Streptococcus thermophilus* counts (6.4 Log CFU/g of yogurt) observed in both yogurts (Figure 3c) is indicative that the viability of lactic acid bacteria is not affected by the supplementation. As represented in Figure 3d, the total polyphenolic content of the control yogurt (11 ± 2.0 mg GAE/100 g of fresh weight yogurt), whose activity can also be attributed to Tyr, Trp and Phe [48], more than doubled with the addition of the extract (29 ± 3.0 mg GAE/100 g of fresh weight yogurt). This represented a higher phenolic content compared to the use of hazelnut skins [49], but inferior when supplementing yogurts with wine grape pomace [47]. These results are attributed to differences at quantitative and qualitative levels of the phenolic structures present in the various agro-industrial by-products. When compared to the total polyphenols added to the yogurt mixture (32.1 mg GAE) with the amount determined in the yogurt mixture after control subtraction, it was observed that only 56% were determined by the Folin–Ciocalteu method, similar to what was observed when protein was added to polyphenols [50]. This variation is possibly attributed to the capability of apple pomace polyphenols to interact with milk proteins, thereby blocking the polyphenolic aromatic rings responsible for their antioxidant properties. Nevertheless, an increase of the antioxidant activity in more than three-fold (from 9 ± 1 to 32 ± 4 mg trolox/100 g of fresh weight yogurt) was measured by the ABTS•$^+$ method (Figure 1e), reflecting the antioxidant properties described for the HWE.

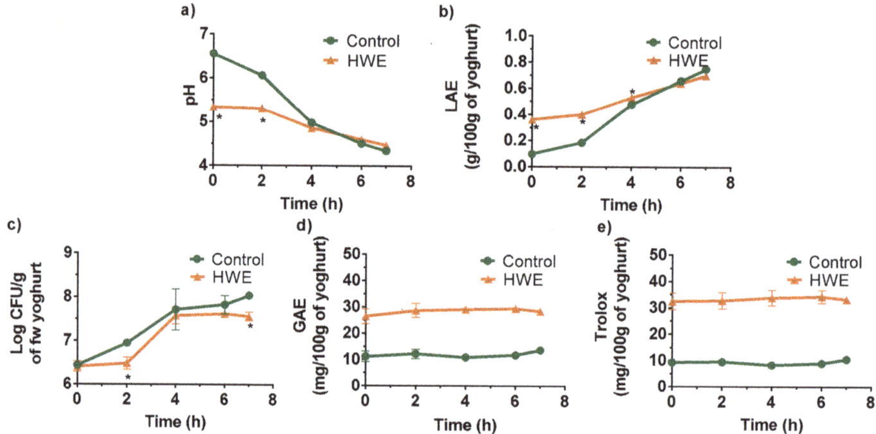

Figure 3. Evolution of (**a**) pH, (**b**) titratable acidity (expressed as lactic acid equivalents (LAE)/100 g of fresh weight yogurt)), (**c**) Streptococcus thermophilus counts (Log CFU/g of fresh weight yogurt), (**d**) total polyphenolic content (mg gallic acid equivalents (GAE)/100 g of fresh weight yogurt), and (**e**) antioxidant activity (mg of Trolox/100 g of fresh weight yogurt) along the fermentation process for the control and supplemented yogurt with the HWE.

As represented in Figure 3a–c the pH decrease, concomitant with the increase of the titratable acidity and *S. thermophilus* counts, demonstrated that the mixtures with and without apple pomace HWE were fermented, yielding a control and a supplemented yogurt. Nevertheless, *S. thermophilus* in the HWE yogurt appeared to have an increased lag phase and reached lower counts at the end of fermentation when compared to the control, which might indicate that the tested concentration may have an inhibitory effect on their growth. Although lactic acid bacteria are generally isolated and remain viable in acidic foods, their growth is inhibited at low pH levels, in particular the growth of *S. thermophilus* [51]. Therefore, it is possible that the extended lag phase observed in the HWE yogurt resulted from its initial acidic pH when compared to the one observed for the control. Nevertheless, the *S. thermophilus* counts exceeded the 7 Log CFU/g, with a final pH below 4.5, which are positive critical factors to inhibit pathogenic microorganisms such as *Listeria monocytogenes*, *Salmonella*, or *Escherichia coli*, and to assure the stability of the product [52,53]. Furthermore, it was observed that the yogurt total polyphenolic content and antioxidant activity (Figure 3d,e) remained unchanged during the fermentation process. This suggested that the polyphenols incorporated into yogurt were not affected. However, given the capability of bacteria to metabolize polyphenols, it is feasible that some apple pomace polyphenolic structures were converted to metabolites that still possess antioxidant properties. These changes might not be reflected on the overall total phenolic content and antioxidant activity, as measured by the Folin–Ciocalteu and ABTS•$^+$ methods. Reports on yogurts supplemented with hazelnut skins [49] and grape pomace [47] suggest that their antioxidant properties can be relatively stable for at least two weeks.

Table 4 shows the nutritional composition, expressed in g/100 g of fresh weight yogurt, and the energetic value, expressed in kcal/100 g of fresh weight yogurt of the control and supplemented yogurt. The nutritional analysis revealed that water was the major component (87%–89%) in both formulations. The addition of the HWE resulted in an increase of the yogurts total sugars from 5.5% to 7.3% and of the reducing sugars from 4.5% to 5.1%. As lactic acid bacteria, particularly *S. thermophilus*, preferentially use lactose over glucose [54], these increments can be attributed to some glucose and fructose and to oligosaccharides and polysaccharides from the HWE that remain after fermentation. From the latter, two distinct groups could be highlighted: (1) high molecular weight (>10 kDa) polysaccharide/polyphenol complexes to which higher short-chain fatty acid production and polyphenol-derived metabolites are attributed than to polyphenols and polysaccharides alone [55], and (2) pectic oligosaccharides to which prebiotic properties are also attributed [56]. It has been shown that these carbohydrate structures also improve yogurt firmness [57]. Protein, fat, and ash corresponded to about 3.2%, 1.5%, and 0.9%, respectively. Overall, the energetic values of both yogurts were 47 and 49 kcal/100 g of fresh weight for the control and HWE yogurt, respectively.

Table 4. Nutritional composition expressed as g/100 g of fresh weight control and supplemented yogurt with apple pomace HWE.

Components	Control	HWE
Moisture	88.8 ± 0.0	87.2 ± 0.1
Total Sugars	5.45 ± 0.12	7.30 ± 0.18
Reducing Sugars	4.49 ± 0.10	5.05 ± 0.04
HWE polysaccharides *		1.43 ± 0.01
Protein	3.32 ± 0.04	3.21 ± 0.08
Fat	1.68 ± 0.15	1.46 ± 0.07
Ash	0.75 ± 0.01	0.84 ± 0.01
Energetic (kcal)	46.5 ± 1.1	49.1 ± 0.7

* Assuming that HWE polysaccharides are preserved during fermentation.

4. Conclusions

In this work, it was shown that by hot water extraction under acidic conditions more than 3 g/kg of dry apple pomace could be obtained. Ultrafiltration demonstrated that in addition to apple native polyphenols, some phenolic structures were probably attached to the high molecular weight material (>10 kDa). Separation of HWE polyphenols by solid-phase extraction allowed us to infer the potential antioxidant and anti-inflammatory capacities, as shown by their scavenging ability towards NO• in chemical and cellular models. When applied to yogurt formulations, apple pomace HWE allowed for achieving a final product with improved fiber content and antioxidant properties when compared to the plain yogurt. However, as this study involves only in vitro assays, its extrapolation to humans cannot be done. This results from the fact that DPPH• and ABTS•$^+$ radicals, although suggestive of antioxidant properties, do not represent physiological radicals while the OH• and NO• scavenging assays were performed using in vitro models. Therefore, these results should be considered as a proof of concept for food chemists and industrials. For the evaluation of possible health benefits, in vivo studies addressing the bioavailability of the polyphenols and the dosage required to observe any antioxidant or anti-inflammatory effects are necessary. In fact, these requirements are regulated by the European Food Safety Authority (EFSA) and must meet not only the requirements for health claims (Regulation (EC) No 1924/2006 of the European Parliament and of the Council), but also for safety (Regulation (EU) No 1169/2011 of the European Parliament and of the Council of 25 October 2011) before any commercialization of a food product stating health benefits. In this context, no statement of the type "antioxidant and anti-inflammatory apple pomace extract/yoghurt" cannot be used at this stage and any commercial exploitation of the developed yoghurt formulation must naturally assure its safety.

Author Contributions: P.A.R.F. contributed to conceptualization, investigation, data curation, and writing the original draft. S.S.F. and R.B. contributed to conceptualization, investigation, data curation, and writing – reviewing and editing. I.F. and M.T.C. contributed to conceptualization, investigation, resources and writing – reviewing and editing. A.P. contributed to conceptualization, investigation, and writing – reviewing and editing. C.P.P. contributed to methodology, resources, and writing – reviewing and editing. E.C. contributed to conceptualization, methodology, supervision, and writing – reviewing and editing. M.A.C. contributed to conceptualization, investigation, methodology, supervision, and writing – reviewing and editing. S.M.C. contributed to conceptualization, investigation, methodology, supervision, and writing – reviewing and editing. D.F.W. contributed to conceptualization, investigation, methodology, supervision, writing – reviewing and editing, and project administration.

Funding: Thanks are due to the University of Aveiro and FCT/MCT for the financial support for the QOPNA research Unit (FCT UID/QUI/00062/2019) and to CITAB research Unit (FCT UID/AGR/04033/2019) through national funds and, where applicable, co-financed by the FEDER, within the PT2020 Partnership Agreement, and to the Portuguese NMR Network. Pedro A. R. Fernandes (SFRH/BD/107731/2015) and Elisabete Coelho (SFRH/BPD/70589/2010) were supported by FCT/MEC grants. Claudia Passos acknowledges FCT (CEECIND/01873/2017) contract. Susana M. Cardoso thanks the research contract under the project AgroForWealth: Biorefining of agricultural and forest by-products and wastes: integrated strategic for valorisation of resources towards society wealth and sustainability (CENTRO-01-0145-FEDER-000001), funded by Centro2020, through FEDER and PT2020. Thanks are due to Indumape for providing the apple pomace and to project Profitapple 3862.

Conflicts of Interest: The authors declare no conflict of interest.

References

1. O'Shea, N.; Arendt, E.K.; Gallagher, E. Dietary fibre and phytochemical characteristics of fruit and vegetable by-products and their recent applications as novel ingredients in food products. *Innov. Food Sci. Emerg. Technol.* **2012**, *16*, 1–10. [CrossRef]

2. Rabetafika, H.N.; Bchir, B.; Blecker, C.; Richel, A. Fractionation of apple by-products as source of new ingredients: Current situation and perspectives. *Trends Food Sci. Technol.* **2014**, *40*, 99–114. [CrossRef]

3. Boyer, J.; Liu, R.H. Apple phytochemicals and their health benefits. *Nutr. J.* **2004**, *3*, 5. [CrossRef] [PubMed]

4. Mirabella, N.; Castellani, V.; Sala, S. Current options for the valorization of food manufacturing waste: A review. *J. Clean. Prod.* **2014**, *65*, 28–41. [CrossRef]

5. Çam, M.; Aaby, K. Optimization of Extraction of Apple Pomace Phenolics with Water by Response Surface Methodology. *J. Agric. Food Chem.* **2010**, *58*, 9103–9111. [CrossRef] [PubMed]

6. Kschonsek, J.; Wolfram, T.; Stöckl, A.; Böhm, V. Polyphenolic Compounds Analysis of Old and New Apple Cultivars and Contribution of Polyphenolic Profile to the In Vitro Antioxidant Capacity. *Antioxidants* **2018**, *7*, 20. [CrossRef] [PubMed]

7. Le Bourvellec, C.; Guyot, S.; Renard, C.M.G.C. Interactions between apple (Malus x domestica Borkh.) polyphenols and cell walls modulate the extractability of polysaccharides. *Carbohydr. Polym.* **2009**, *75*, 251–261. [CrossRef]

8. Fernandes, P.A.R.; Le Bourvellec, C.; Renard, C.M.G.C.; Nunes, F.M.; Bastos, R.; Coelho, E.; Wessel, D.F.; Coimbra, M.A.; Cardoso, S.M. Revisiting the chemistry of apple pomace polyphenols. *Food Chem.* **2019**. [CrossRef] [PubMed]

9. Suárez, B.; Álvarez, Á.L.; García, Y.D.; del Barrio, G.; Lobo, A.P.; Parra, F. Phenolic profiles, antioxidant activity and in vitro antiviral properties of apple pomace. *Food Chem.* **2010**, *120*, 339–342. [CrossRef]

10. Kennedy, M.; List, D.; Lu, Y.; Foo, L.Y.; Newman, R.H.; Sims, I.M.; Bain, P.J.S.; Hamilton, B.; Fenton, G. Apple Pomace and Products Derived from Apple Pomace: Uses, Composition and Analysis. In *Analysis of Plant Waste Materials*; Linskens, H.F., Jackson, J.F., Eds.; Springer: Berlin/Heidelberg, Germany, 1999; pp. 75–119.

11. Rohn, S.; Buchner, N.; Driemel, G.; Rauser, M.; Kroh, L.W. Thermal Degradation of Onion Quercetin Glucosides under Roasting Conditions. *J. Agric. Food Chem.* **2007**, *55*, 1568–1573. [CrossRef]

12. Le Bourvellec, C.; Gouble, B.; Bureau, S.; Loonis, M.; Plé, Y.; Renard, C.M.G.C. Pink Discoloration of Canned Pears: Role of Procyanidin Chemical Depolymerization and Procyanidin/Cell Wall Interactions. *J. Agric. Food Chem.* **2013**, *61*, 6679–6692. [CrossRef] [PubMed]

13. Passos, C.P.; Ferreira, S.S.; Serôdio, A.; Basil, E.; Marková, L.; Kukurová, K.; Ciesarová, Z.; Coimbra, M.A. Pectic polysaccharides as an acrylamide mitigation strategy—Competition between reducing sugars and sugar acids. *Food Hydrocoll.* **2018**, *81*, 113–119. [CrossRef]

14. Sosulski, F.W.; Imafidon, G.I. Amino acid composition and nitrogen-to-protein conversion factors for animal and plant foods. *J. Agric. Food Chem.* **1990**, *38*, 1351–1356. [CrossRef]

15. Blumenkrantz, N.; Asboe-Hansen, G. New method for quantitative determination of uronic acids. *Anal. Biochem.* **1973**, *54*, 484–489. [CrossRef]

16. Brunton, N.P.; Gormley, T.R.; Murray, B. Use of the alditol acetate derivatisation for the analysis of reducing sugars in potato tubers. *Food Chem.* **2007**, *104*, 398–402. [CrossRef]

17. Singleton, V.L.; Rossi, J.A. Colorimetry of Total Phenolics with Phosphomolybdic-Phosphotungstic Acid Reagents. *Am. J. Enol. Viticult.* **1965**, *16*, 144.

18. Brand-Williams, W.; Cuvelier, M.E.; Berset, C. Use of a free radical method to evaluate antioxidant activity. *LWT Food Sci. Technol.* **1995**, *28*, 25–30. [CrossRef]

19. Re, R.; Pellegrini, N.; Proteggente, A.; Pannala, A.; Yang, M.; Rice-Evans, C. Antioxidant activity applying an improved ABTS radical cation decolorization assay. *Free Radic. Biol. Med.* **1999**, *26*, 1231–1237. [CrossRef]

20. Kunchandy, E.; Rao, M.N.A. Oxygen radical scavenging activity of curcumin. *Int. J. Pharm.* **1990**, *58*, 237–240. [CrossRef]

21. Catarino, M.D.; Silva, A.M.S.; Cruz, M.T.; Cardoso, S.M. Antioxidant and anti-inflammatory activities of *Geranium robertianum* L. decoctions. *Food Funct.* **2017**, *8*, 3355–3365. [CrossRef]

22. Búfalo, M.C.; Ferreira, I.; Costa, G.; Francisco, V.; Liberal, J.; Cruz, M.T.; Lopes, M.C.; Batista, M.T.; Sforcin, J.M. Propolis and its constituent caffeic acid suppress LPS-stimulated pro-inflammatory response by blocking NF-κB and MAPK activation in macrophages. *J. Ethnopharmacol.* **2013**, *149*, 84–92. [CrossRef] [PubMed]

23. Renard, C.M.G.C.; Lemeunier, C.; Thibault, J.F. Alkaline extraction of xyloglucan from depectinised apple pomace: Optimisation and characterisation. *Carbohydr. Polym.* **1995**, *28*, 209–216. [CrossRef]

24. DuBois, M.; Gilles, K.A.; Hamilton, J.K.; Rebers, P.t.; Smith, F. Colorimetric method for determination of sugars and related substances. *Anal. Chem.* **1956**, *28*, 350–356. [CrossRef]

25. Miller, G.L. Use of dinitrosalicylic acid reagent for determination of reducing sugar. *Anal. chem.* **1959**, *31*, 426–428. [CrossRef]

26. European Parliament. The Council of the European Union. Regulation (EU) No 1169/2011. *Off. J. Eur. Union* **2011**, *L 304*, 18–63.

27. Percy, A.E.; Melton, L.D.; Jameson, P.E. Xyloglucan and hemicelluloses in the cell wall during apple fruit development and ripening. *Plant Sci.* **1997**, *125*, 31–39. [CrossRef]

28. Renard, C.M.G.C.; Voragen, A.G.J.; Thibault, J.F.; Pilnik, W. Studies on apple protopectin V: Structural studies on enzymatically extracted pectins. *Carbohydr. Polym.* **1991**, *16*, 137–154. [CrossRef]

29. Ferreira, D.; Guyot, S.; Marnet, N.; Delgadillo, I.; Renard, C.M.G.C.; Coimbra, M.A. Composition of Phenolic Compounds in a Portuguese Pear (*Pyrus communis* L. Var. S. Bartolomeu) and Changes after Sun-Drying. *J. Agric. Food Chem.* **2002**, *50*, 4537–4544. [CrossRef] [PubMed]

30. Reis, S.F.; Rai, D.K.; Abu-Ghannam, N. Water at room temperature as a solvent for the extraction of apple pomace phenolic compounds. *Food Chem.* **2012**, *135*, 1991–1998. [CrossRef]

31. Guyot, S.; Marnet, N.; Sanoner, P.; Drilleau, J.-F. Variability of the Polyphenolic Composition of Cider Apple (Malus domestica) Fruits and Juices. *J. Agric. Food Chem.* **2003**, *51*, 6240–6247. [CrossRef]

32. Renard, C.M.G.C.; Baron, A.; Guyot, S.; Drilleau, J.F. Interactions between apple cell walls and native apple polyphenols: Quantification and some consequences. *Int. J. Biol. Macromol.* **2001**, *29*, 115–125. [CrossRef]

33. Guyot, S.; Le Bourvellec, C.; Marnet, N.; Drilleau, J.F. Procyanidins are the most Abundant Polyphenols in Dessert Apples at Maturity. *LWT Food Sci. Technol.* **2002**, *35*, 289–291. [CrossRef]

34. Gonçalves, F.J.; Rocha, S.M.; Coimbra, M.A. Study of the retention capacity of anthocyanins by wine polymeric material. *Food Chem.* **2012**, *134*, 957–963. [CrossRef] [PubMed]

35. Gonçalves, F.J.; Fernandes, P.A.R.; Wessel, D.F.; Cardoso, S.M.; Rocha, S.M.; Coimbra, M.A. Interaction of wine mannoproteins and arabinogalactans with anthocyanins. *Food Chem.* **2018**, *243*, 1–10. [CrossRef] [PubMed]

36. Cruz, M.G.; Bastos, R.; Pinto, M.; Ferreira, J.M.; Santos, J.F.; Wessel, D.F.; Coelho, E.; Coimbra, M.A. Waste mitigation: From an effluent of apple juice concentrate industry to a valuable ingredient for food and feed applications. *J. Clean. Prod.* **2018**. [CrossRef]

37. Sánchez-Rabaneda, F.; Jáuregui, O.; Lamuela-Raventós, R.M.; Viladomat, F.; Bastida, J.; Codina, C. Qualitative analysis of phenolic compounds in apple pomace using liquid chromatography coupled to mass spectrometry in tandem mode. *Rapid Commun. Mass Spectrom.* **2004**, *18*, 553–563. [CrossRef]

38. Millet, M.; Poupard, P.; Le Quéré, J.-M.; Bauduin, R.; Guyot, S. Haze in Apple-Based Beverages: Detailed Polyphenol, Polysaccharide, Protein, and Mineral Compositions. *J. Agric. Food Chem.* **2017**, *65*, 6404–6414. [CrossRef]

39. Nicolas, J.J.; Richard-Forget, F.C.; Goupy, P.M.; Amiot, M.J.; Aubert, S.Y. Enzymatic browning reactions in apple and apple products. *Crit. Rev. Food Sci. Nutr.* **1994**, *34*, 109–157. [CrossRef]

40. Bernillon, S.; Guyot, S.; Renard, C.M.G.C. Detection of phenolic oxidation products in cider apple juice by high-performance liquid chromatography electrospray ionisation ion trap mass spectrometry. *Rapid Commun. Mass Spectrom.* **2004**, *18*, 939–943. [CrossRef]

41. Mouls, L.; Fulcrand, H. Identification of new oxidation markers of grape-condensed tannins by UPLC–MS analysis after chemical depolymerization. *Tetrahedron* **2015**, *71*, 3012–3019. [CrossRef]

42. Schieber, M.; Chandel, N.S. ROS Function in Redox Signaling and Oxidative Stress. *Curr. Biol.* **2014**, *24*, 453–462. [CrossRef] [PubMed]

43. Fernandez-Panchon, M.S.; Villano, D.; Troncoso, A.M.; Garcia-Parrilla, M.C. Antioxidant Activity of Phenolic Compounds: From In Vitro Results to In Vivo Evidence. *Crit. Rev. Food Sci. Nutr.* **2008**, *48*, 649–671. [CrossRef] [PubMed]

44. Mittal, M.; Siddiqui, M.R.; Tran, K.; Reddy, S.P.; Malik, A.B. Reactive Oxygen Species in Inflammation and Tissue Injury. *Antioxid. Redox Signal.* **2014**, *20*, 1126–1167. [CrossRef] [PubMed]

45. Chen, Y.-C.; Shen, S.-C.; Lee, W.-R.; Hou, W.-C.; Yang, L.-L.; Lee, T.J.F. Inhibition of nitric oxide synthase inhibitors and lipopolysaccharide induced inducible NOS and cyclooxygenase-2 gene expressions by rutin, quercetin, and quercetin pentaacetate in RAW 264.7 macrophages. *J. Cell. Biochem.* **2001**, *82*, 537–548. [CrossRef] [PubMed]

46. Huang, W.-C.; Chang, W.-T.; Wu, S.-J.; Xu, P.-Y.; Ting, N.-C.; Liou, C.-J. Phloretin and phlorizin promote lipolysis and inhibit inflammation in mouse 3T3-L1 cells and in macrophage-adipocyte co-cultures. *Mol. Nutr. Food Res.* **2013**, *57*, 1803–1813. [CrossRef] [PubMed]

47. Tseng, A.; Zhao, Y. Wine grape pomace as antioxidant dietary fibre for enhancing nutritional value and improving storability of yogurt and salad dressing. *Food Chem.* **2013**, *138*, 356–365. [CrossRef] [PubMed]

48. Shah, N.P. Effects of milk-derived bioactives: An overview. *Br. J. Nutr.* **2000**, *84*, 3–10. [CrossRef]

49. Bertolino, M.; Belviso, S.; Dal Bello, B.; Ghirardello, D.; Giordano, M.; Rolle, L.; Gerbi, V.; Zeppa, G. Influence of the addition of different hazelnut skins on the physicochemical, antioxidant, polyphenol and sensory properties of yogurt. *LWT Food Sci. Technol.* **2015**, *63*, 1145–1154. [CrossRef]

50. Xiao, J.B.; Huo, J.L.; Yang, F.; Chen, X.Q. Noncovalent Interaction of Dietary Polyphenols with Bovine Hemoglobin in Vitro: Molecular Structure/Property–Affinity Relationship Aspects. *J. Agric. Food Chem.* **2011**, *59*, 8484–8490. [CrossRef]

51. Hutkins, R.W.; Nannen, N.L. pH Homeostasis in Lactic Acid Bacteria1. *J. Dairy Sci.* **1993**, *76*, 2354–2365. [CrossRef]

52. Massa, S.; Altieri, C.; Quaranta, V.; De Pace, R. Survival of *Escherichia coli* O157:H7 in yogurt during preparation and storage at 4 °C. *Lett. Appl. Microbiol.* **1997**, *24*, 347–350. [CrossRef] [PubMed]

53. Massa, S.; Trovatelli, L.D.; Canganella, F. Survival of *Listeria monocytogenes* in yogurt during storage at 4 °C. *Lett. Appl. Microbiol.* **1991**, *13*, 112–114. [CrossRef]

54. Bruckner, R.; Titgemeyer, F. Carbon catabolite repression in bacteria: Choice of the carbon source and autoregulatory limitation of sugar utilization. *FEMS Microbiol. Lett.* **2002**, *209*, 141–148. [CrossRef]

55. Pérez-Jiménez, J.; Díaz-Rubio, M.E.; Saura-Calixto, F. Non-extractable polyphenols, a major dietary antioxidant: Occurrence, metabolic fate and health effects. *Nutr. Res. Rev.* **2013**, *26*, 118–129. [CrossRef] [PubMed]

56. Ferreira-Lazarte, A.; Kachrimanidou, V.; Villamiel, M.; Rastall, R.A.; Moreno, F.J. In vitro fermentation properties of pectins and enzymatic-modified pectins obtained from different renewable bioresources. *Carbohydr. Polym.* **2018**, *199*, 482–491. [CrossRef]

57. Wang, X.; Kristo, E.; LaPointe, G. The effect of apple pomace on the texture, rheology and microstructure of set type yogurt. *Food Hydrocoll.* **2019**, *91*, 83–91. [CrossRef]

Article

Towards a Zero-Waste Biorefinery Using Edible Oils as Solvents for the Green Extraction of Volatile and Non-Volatile Bioactive Compounds from Rosemary

Ying Li [1,2,*], Kunnitee Bundeesomchok [2,3], Njara Rakotomanomana [2], Anne-Sylvie Fabiano-Tixier [2], Romain Bott [4], Yong Wang [1] and Farid Chemat [2,*]

1 JNU-UPM International Joint Laboratory on Plant Oil Processing and Safety, Guangdong Engineering Technology Research Center for Cereal and Oil Byproduct Biorefinery, Department of Food Science and Engineering, Jinan University, Guangzhou 510632, China; twyong@jnu.edu.cn
2 GREEN Extraction Team, Université d'Avignon et des Pays de Vaucluse, INRA, UMR408, F-84000 Avignon, France; purepharm@hotmail.com (K.B.); njara.rakotomanomana@univ-avignon.fr (N.R.); anne-sylvie.fabiano@univ-avignon.fr (A.-S.F.-T.)
3 Department of Pharmacognosy and Pharmaceutical Botany, Faculty of Pharmaceutical Sciences, Prince of Songkla University, Hat-Yai, Songkhla 90112, Thailand
4 INRA, UMR408, Securité et Qualité des Produits d'Origine Végétale, F-84000 Avignon, France; Romain.Bott@paca.inra.fr
* Correspondence: yingli@jnu.edu.cn (Y.L.); farid.chemat@univ-avignon.fr (F.C.); Tel.: +86-20-85223429 (Y.L.); +33-490144465 (F.C.)

Received: 6 May 2019; Accepted: 18 May 2019; Published: 21 May 2019

Abstract: The zero-waste biorefinery concept inspired a green oleo-extraction of both natural volatile (e.g., borneol, camphor, o-cymene, eucalyptol, limonene, α-pinene, and terpinen-4-ol) and non-volatile (e.g., carnosol, carnosic, and rosmarinic acid) bioactive compounds from rosemary leaves with vegetable oils and their amphiphilic derivatives as simple food-grade solvents. It is noteworthy that soybean oil could obtain the highest total phenolic compounds (TPCs) among 12 refined oils including grapeseed, rapeseed, peanut, sunflower, olive, avocado, almond, apricot, corn, wheat germ, and hazelnut oils. Furthermore, the addition of oil derivatives to soybean oils, such as glyceryl monooleate (GMO), glyceryl monostearate (GMS), diglycerides, and soy lecithin in particular, could not only significantly enhance the oleo-extraction of non-volatile antioxidants by 66.7% approximately, but also help to remarkably improve the solvation of volatile aroma compounds (VACs) by 16% in refined soybean oils. These experimental results were in good consistency with their relative solubilities predicted by the more sophisticated COSMO-RS (COnductor like Screening MOdel for Real Solvents) simulation. This simple procedure of using vegetable oils and their derivatives as bio-based solvents for simultaneously improving the extraction yield of natural antioxidants and flavors from rosemary showed its great potential in up-scaling with the integration of green techniques (ultrasound, microwave, etc.) for zero-waste biorefinery from biomass waste to high value-added extracts in future functional food and cosmetic applications.

Keywords: green oleo-extraction; zero-waste biorefinery; natural antioxidants and flavors; food-grade solvents; vegetable oils and derivatives; relative solubility simulation

1. Introduction

The large quantities of biomass wastes generated from food processing industries worldwide could be considered as a huge potential resource with high-value compounds for biorefinery, which can provide bio-based chemicals and renewable energy with high added value rather than environmental pollution compromising the economy, environment, and human society in the future [1,2]. Although the existing biorefinery technologies (anaerobic digestion, fermentation, hydrothermal conversion, pyrolysis, etc.) have been widely studied and applied in biomass waste reutilization, increasing concern about high setup costs accompanied by incomplete resource utilization, and even secondary pollution, is compelling the scientific, industrial, and government communities to focus more efforts on developing green biorefinery technologies that integrate innovative technologies, which is crucial in achieving maximum recognition of biowastes as chemical and energy resources [3].

Extraction, as the key step in many common process industries and the major expenditure ($\geq$50%) of the total costs, has been paramount within the scope of biorefinery technological development in particular [4]. Considering the increasing demand to reduce the amount of time, energy, costs, and solvents involved in conventional extraction procedures, green extraction with its principles regarding renewable raw materials, alternative solvents [5,6] reduced energy consumption, process intensification, and production of co-products instead of wastes may be well suited as an alternative for a green biorefinery [7]. Currently, several representative technologies such as ultrasound, microwave, supercritical fluids, and instantaneous controlled decompression have successfully proved their effectiveness for the green extraction of natural products from laboratorial to industrial scale [8–12]. However, only a few studies had achieved a biowaste biorefinery according to the green extraction concept and principles.

Plants, especially herbs and spices, have been the major sources for the extraction of numerous natural products with bioactive properties [13]. The reference matrix chosen for this study is rosemary, a well-known ornamental and aromatic plant that has been cultivated along the Mediterranean Sea and widely applied for different purposes. It contains phenolic compounds of great interest for their high antioxidant activities, which are attributed to known carnosol, carnosic, and rosmarinic acids as main constituents [14–16]. After the production of rosemary essential oils, the natural phenolic compounds in the residual rosemary have still gained growing interest for direct extraction. Although water or its ethanolic mixture have been recognized as green solvents for extracting rosemary in a sustainable and safe manner, the low extraction yield and unsuitability for the extraction of maximal bioactive compounds other than a single volatile or non-volatile compound limit its application [17,18]. Recently, the application of vegetable oils as alternative solvents for extracting bioactive compounds in various fields has been revisited. These non-polar edible solvents have the advantage of being non-toxic, non-volatile, and non-irritating. Moreover, the endogenous micronutrients like phospholipids, sterols, monoglycerides, and diglycerides are not handicaps for food, cosmetic, or nutraceutical applications [19]. Considering the polarity of oil constituents and the previous experience, it is believed that such bio-based solvents could also be used for the total valorization of rosemary based on the biorefinery and green extraction concept [20–22].

The novelty of this work is the development of a direct oleo-extraction method towards the zero-waste biorefinery concept using vegetable oils and their amphiphilic derivatives as solvents, which facilitates the simultaneous extraction of both volatile and non-volatile bioactive compounds from rosemary leaves (Figure 1). The large consumption of conventional volatile organic solvents was replaced in this maximal extraction of rosemary compounds so that potential risk factors could be avoided. Moreover, the COSMO-RS (COnductor like Screening MOdel for Real Solvents) simulation study helped to theoretically predict the solvent–solute miscibility and further confirm the experimental extractability of all oily solvents for a better comprehension of the dissolving mechanism. In addition, the maximal extraction efficiency of refined soybean oils was verified for the first time with the addition of their amphiphilic derivatives as food-grade surfactants in both a theoretical and an experimental way.

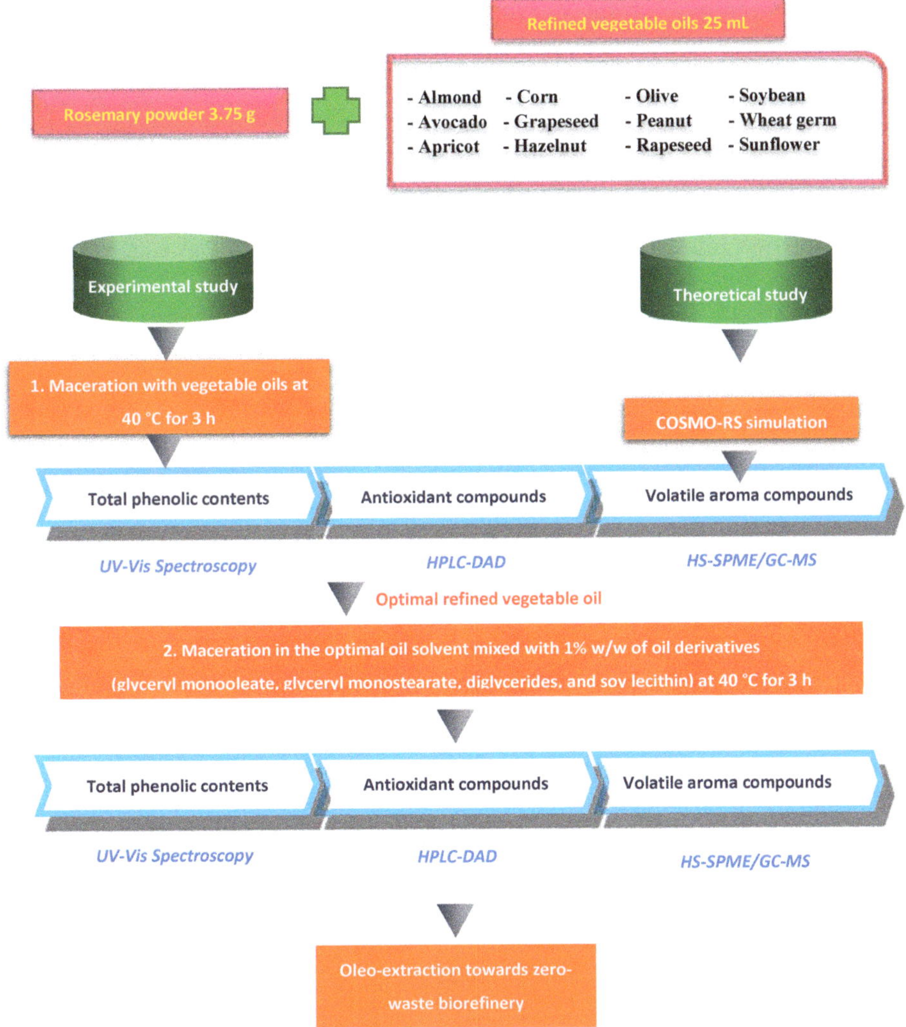

Figure 1. Schematic experimental design. UV-Vis: Ultraviolet-visible; HPLC-DAD: High performance liquid chromatography- diode array detector; HS-SPME: Head space solid-phase microextraction; GC-MS: Gas chromatography-mass spectroscopy

2. Materials and Methods

2.1. Plant Material and Chemicals

Dried rosemary (*Rosmarinus officinalis* L.) leaves after deodorization were provided by Naturex, Avignon, France. For extraction solvents, refined corn, olive, avocado, wheat germ, hazelnut, apricot, sweet almond, and soybean oils were obtained from *ieS* LABO, Oraison, France. Refined peanut, sunflower, and rapeseed oils were obtained from Auchan, Avignon, France. Refined grapeseed oil was obtained from Tramier, Marseille, France. Glyceryl monooleate (GMO), glyceryl monostearate (GMS), diglycerides (Geleol™, Gattefosse), and soy lecithin were obtained from Oleos, Lunel, France. For analysis, solvents of analytical grade including methanol (99.8%), acetonitrile (99.8%), n-hexane (99.5%), phosphoric acid (85%), trifluoroacetic acid (99.8%), sodium bicarbonate (99.7%), and Folin–Ciocalteu's reagent, and water of HPLC (high-performance liquid chromatography) grade as well, were purchased from VWR Chemicals, Darmstadt, Germany. Standards of carnosol (purity 98%), carnosic, and rosmarinic acid (purity 97%) used in the HPLC analysis were purchased from Sigma-Aldrich, Munich, Germany.

2.2. Fatty Acid Methyl Ester (FAME) Analysis

The fatty acid composition of various vegetable oil solvents was determined according to our developed FAME method [23]. FAMEs were obtained after the methylation of oil samples, from which the supernatants were transferred into vials for further GC-FID (Gas chromatography- flame ionization detector) analysis by a 7820A GC system (Agilent Technologies, Palo Alto, CA, USA) equipped with an FID detector and auto-sampler using a BD-EN14103 capillary column (30 m × 0.32 mm × 0.25 μm) with helium as a carrier gas at the linear velocity of 35 cm/s and glyceryl triheptadecanoate ($C_{54}H_{104}O_6$) as the internal standard. One microliter of sample was injected in split mode (split ratio 1:20) at 250 °C. The oven temperature program was operated as follows: initial temperature at 50 °C, increasing at 20 °C/min to 180 °C and at 2 °C/min from 180 °C to 230 °C, held isothermally at 230 °C for 10 min. Data were collected using Agilent EZChrom Elite (Palo Alto, CA, USA) software and fatty acids were identified by referring to 37 FAME standards (Supelco, Bellefonte, PA, USA). FAMEs were quantified as the relative percentage of the total fatty acids.

2.3. Solid–Liquid Extraction

Solid–liquid extractions were referred to a previous study [6]. Briefly, the dried rosemary leaves were mechanically ground and passed through No. 60 mesh screens (0.25 mm) for further maceration. The resulting fine powders (3.75 g) were poured into flasks containing 25 mL of various vegetable oils on a magnetic stirrer plate (RT-10, IKAMAG, Staufen, Germany) at 40 °C for 3 h. Meanwhile, the total phenolic content (TPC) in the rosemary leaves was determined by reflux extraction (EMA0250 Thermo Fisher Scientific, Waltham, MA, USA) of rosemary leaf powders (3.75 g) in 25 mL of aqueous methanol (methanol/water, 90:10 *v/v*) at the boiling point for 30 min. All samples were then centrifuged at 2739× *g* for 15 min at 4 °C in a refrigerated centrifuge (4–16k, Sigma-Aldrich, Munich, Germany). A modified liquid–liquid extraction procedure was used to obtain the phenolic extracts in vegetable oils, where 10 mL of vegetable oils were mixed with n-hexane (1:1, *v/v*) and then extracted with 20 mL of aqueous methanol (methanol/water, 60:40 *v/v*) thrice. The extracts were combined, washed with n-hexane, and then filtered through filters (0.45 μm). All experiments were carried out in triplicate. The methanolic extracts were evaporated to 5 mL and stored at −18 °C for subsequent analyses.

2.4. Folin–Ciocalteu Assay

Folin–Ciocalteu assay was used to determine the TPC in the oily extracts. Folin–Ciocalteu's reagent in 500 μL of water (20%, *v/v*) was mixed with 50 μL of methanolic extracts before adding 1 mL of $NaHCO_3$ solution (10%) and then placed in the dark for 30 min. The absorbance was measured at 760 nm against the blank using a UV-Vis spectrometer (Biochrom Libra S22, Cambridge, UK). The TPC measurements were performed in triplicate and the results were reported as milligrams of rosmarinic acid equivalent per 3.75 g of dried materials.

2.5. HPLC Analysis

Carnosol, carnosic, and rosmarinic acids were quantitatively analyzed using an HPLC system (Agilent 1100, Les Ulis, France) equipped with a photo diode array detector (DAD) according to our internal method developed [9], whose procedures are detailed below. Carnosol and carnosic acid were detected at a wavelength of 230 nm using a C18 column (1.8 μm, 4.6 mm × 50 mm, Zorbax Eclipse XBD-C18, Agilent Technologies, Courtaboeuf, France). The mobile phase was isocratic and composed of 0.5% phosphoric acid in water:acetonitrile (35:65, *v/v*). Samples (5 μL) were injected with a flow rate of 1.5 mL/min and the column oven temperature was 25 °C. Rosmarinic acid was detected at a wavelength of 328 nm using a C18 column (5 μm, 4.6 mm × 250 mm, Zorbax SB, Agilent Technologies, Courtaboeuf, France). The mobile phase composed of acetonitrile (32%) and 0.1% of trifluoroacetic acid in water (68%). Samples (5 μL) were injected with a flow rate of 1 mL/min and the column oven temperature was 20 °C with a 10 min runtime.

2.6. Headspace Volatile Analysis

The volatile aroma compounds (VACs) in oily rosemary extracts were identified and quantified by headspace solid-phase microextraction (HS-SPME) coupled to gas chromatography mass spectrometry (GC/MS-QP2010, Shimadzu, Kyoto, Japan) with an AOC 5000 auto-injector (Shimadzu, Kyoto, Japan) [24]. The auto sampler was operated in SPME mode using a divinylbenzene-carboxen-polydimethylsiloxane fiber (2 cm, 23 gauge, 50/30 μm DVB/CAR/PDMS; Supelco, Bellefonte, PA, USA) for extraction. For each extraction, oily extract (4 g) was hermetically sealed in screw-top vials (20 mL) containing aluminum seals and PTFE/silicone septa from Grace, France. The samples were equilibrated during incubation time at 35 °C for 15 min before the automatic insertion of the SPME device into the sealed vial, where the fiber was exposed to the headspace of samples at the same temperature for 25 min. The samples were agitated during the incubation and extraction procedures, and the SPME fiber was subsequently removed and inserted into the GC-MS (Gas chromatography-mass spectroscopy) injector port for desorption at 250 °C for 5 min.

The GC-MS analysis was performed using a QP2010 (Shimadzu, Kyoto, Japan) equipped with a capillary column UB-WAX (30 m × 0.25 mm × 0.5 μm) and helium as a carrier gas at the constant flow of 35 cm/s. The initial oven temperature of 35 °C was held for 2 min, then rose at 5 °C/min until 230 °C. The temperature of the transfer line connecting the GC-MS was held at 250 °C. The inlet was operated in the splitless mode and the mass spectrometer operated in the electron impact mode at 70 eV with continuous scans (every 0.2 s) over the mass-to-charge ratio (m/z) from 35 to 250. Data were collected using GCMSsolution software 2.40 (Shimadzu, Kyoto, Japan) and the major VACs were identified as compared to their linear retention index and mass spectra with those of authentic standards, as well as with the NIST'98 (US National Institute of Standards and Technology (NIST), Gaithersburg, MD, USA) mass spectral database.

2.7. In Silico Solubility Study: COSMO-RS

The in silico solubility of main antioxidant compounds (carnosol, carnosic, and rosmarinic acid) and VACs (eucalyptol, camphor, α-pinene, borneol, *o*-cymene, limonene, and terpinen-4-ol) from rosemary in the mixture of optimal oil solvent and its derivatives was studied by the more sophisticated COSMO-RS software package from COSMO*logic*, Leverkusen, Germany. The chemical structure of all solvents and solutes studied was mutually transformed by JChemPaint version 3.3 (GitHub Pages, San Francisco, CA, USA) into the simplified molecular input line entry syntax (SMILES) notations, which could be further simulated to their three-dimensional σ-surface modelling (Figure S1, Supplementary Materials) by embedded Turbomole (TmoleX, version 7.1), where green to yellow codes the weakly polar surfaces, blue represents electron-deficient regions (δ^+), and red codes electron-rich regions (δ^-). This 3D information on the repartition of charge density on the molecular surface can be reduced to σ-profiles and σ-potentials, which could be used for subsequent prediction of the interactions between oily solvents and bioactive compounds based on a quantum-chemical approach. COSMO-RS is known as a modern and powerful method with the combination of quantum chemical considerations (COSMO) and statistical thermodynamics (RS) for molecular description with predicted thermodynamic properties and further solvent prescreening without any experimental data [18].

As depicted in Figure 2, COSMO-RS generally involves two major steps. All molecules are embedded into virtual conductors simulated in the first step by the COSMO model, where the molecule induces a polarization charge density (σ) on its surface that is a good local descriptor of the molecular surface polarity. Therefore, molecules during the quantum calculation are converged to their energetically optimal states in the conductor with respect to electron density and geometry. Density functional theory (DFT) with triple zeta valence polarized basis set (TZVP) were used as the standard quantum chemical method throughout this study.

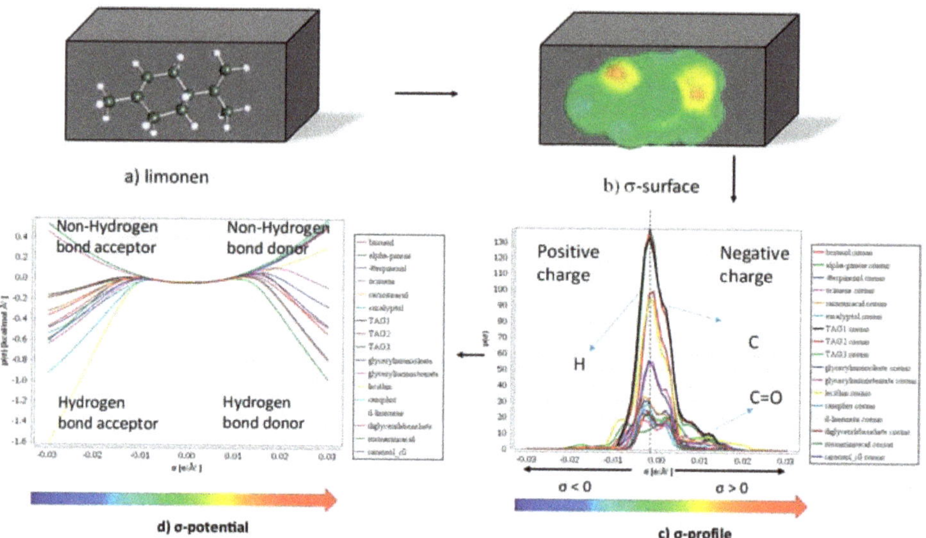

Figure 2. Schematic COSMO-RS (COnductor like Screening MOdel for Real Solvents) step-wise procedures.

The statistical thermodynamics calculation was used in the second step. This polarization charge density was used for the quantification of the interaction energy of pair-wise interacting surface segments concerning electrostatics and hydrogen bonding. The 3D distribution of the polarization charges on the surface of each molecule was converted into a surface composition function (σ-profile) that provided the information about molecular polarity distribution. The thermodynamics of molecular interactions based on the σ-profile obtained was then calculated to chemical potential of the surface segment (σ-potential) using the COSMO*thermX* program (version C30 release 13.01). The σ-potential describes the likeliness of the solute compound interacting with solvents according to their polarities and hydrogen bonds, where the part of the negative charge of molecules (hydrogen bond acceptor) was located on the right side with positive σ-profile values, whereas the part of the positive charge (hydrogen bond donor) was located on the left side with negative σ-profile values. The solvent's affinity for polarity surface could be interpreted according to the σ-profile and σ-potential in order to better understand the solute–solvent interaction in a mixture state.

In addition, COSMO*thermX* also calculated the relative solubility of COSMO-RS simulated volatile and non-volatile bioactive compounds in various oil solvents and mixtures with their derivatives based on the logarithm of the solubility in mole fractions. The relative solubility x_j of compound j is always calculated in infinite dilution according to the equation as follows:

$$\log_{10}(x_j) = \log_{10}\left[\frac{exp(\mu_j^{pure} - \mu_j^{solvent} - \Delta G_{j.fusion})}{RT}\right], \tag{1}$$

where μ_j^{pure} is the chemical potential of pure compound j (J/mol), $\mu_j^{solvent}$ is the chemical potential of j at infinite dilution (J/mol), $\Delta G_{j.fusion}$ is the free energy of fusion of j (J/mol), x_j is the solubility of j (g/g solvent), R is the gas constant, and T is the temperature (K).

The logarithm of the best solubility is set to 0, and all other solvents were given referring to the best solvent. The solubility calculation of major antioxidants and VACs in vegetable oils and mixtures with their derivatives was all performed at 40 °C. A solvent with a $\log10(x_j)$ value of -1.00 yields a solubility which is decreased by a factor 10 compared to the best solvent.

2.8. Statistical Analysis

The experimental solubility of bioactive compounds extracted in 16 oily solvents was further transformed using MATLAB 2015a (The MathWorks, Inc., Natick, MA, USA). First, principal component analysis (PCA) reduced the dimensionality of our data set (individuals: oily solvents; variables: TPC, antioxidants, and VACs) by linear combination into new coordinate systems. The solute concentrations in 16 oily solvent systems were taken for the principal component determination so as to compare the solubility of these 16 solvents in PCA plotting graphs. Clustering analysis was used to classify the closest individuals into clusters, where Ward's hierarchical clustering was applied to calculate dissimilarities from the Euclidean distances and aggregation criterion corresponding to the minimization of the within-cluster inertia and the maximization of between-cluster inertia. Concerning the extractability of non-volatile and volatile bioactive compounds in different solvents, this statistical method could partition oily solvents into homogeneous clusters with a low within-variability, which are different from others with a high between-variability. Given this, a dendrogram was made to illustrate the aggregations made at each successive stage.

3. Results

3.1. Composition of Vegetable Oils

As shown in Table 1, the fatty acid composition of different vegetable oils was affected by the nature and origin [16]. According to the main three fatty acid chains on the triglyceride (TAG), it was noted that most vegetable oils (rapeseed, hazelnut, sweet almond, apricot, sunflower, and peanut oils) contained the same major fatty acids (e.g., oleic acid–oleic acid–linoleic acid (OOL)) with different proportions. Similarly, the major fatty acid compositions in grapeseed, corn, wheat germ, and soybean oils were oleic acid–linoleic acid–linoleic acid (OLL) and avocado oil had its own main fatty acid composition as oleic acid–oleic acid–palmitic acid (OOP). Moreover, three oleic acids (OOO) was inferred in the TAG of olive oil. The above-mentioned four categories of three major fatty acids represented 12 vegetable oils, which could be selected to constitute TAG structures for further computational simulations. Apart from the major composition in oils, it must be mentioned that minor compounds including diglycerides (DAGs), monoglycerides (MAGs), and phospholipids could be considered as food-grade surfactants, which can strongly affect the physiochemical and dissolving properties of the vegetable oils [25].

Table 1. Composition of fatty acids in vegetable oils identified by fatty acid methyl ester (FAME) analysis.

Fatty Acid Composition (%) [a]	Refined Vegetable Oils											
	OOP	OOO	OOL	OOL	OOL	OOL	OOL	OOL	OLL	OLL	OLL	OLL
	Avocado	Olive	Rape Seed	Hazel Nut	Almond	Apricot	Sunflower	Peanut	Grapeseed	Corn	Wheat Germ	Soybean
Palmitic acid (C16:0)	16.5	10.4	4.1	6.2	5.1	5.3	4.4	6.0	6.0	11.1	11.8	10.7
Palmitoleic acid (C16:1)	8.5	0.8	0.2	0.4	0.7	0.8	0.1	0.1	0.1	-	0.2	0.5
Stearic acid (C18:0)	0.6	3.5	-	2.0	2.1	1.1	-	-	3.3	1.6	2.6	2.8
Oleic acid (C18:1)	63.9	78.3	62.5	74.2	66.5	59.8	55.5	73.2	14.4	29.5	29.4	23.2
Linoleic acid (C18:2)	9.5	5.5	19.8	16.4	24.7	28.9	35.3	11.9	75.6	55.8	52.0	55.6
α-Linolenic acid (C18:3)	0.6	0.6	9.7	0.1	0.1	0.1	0.1	0.1	0.1	0.9	-	6.3
Arachidic acid (C20:0)	0.1	0.4	0.3	0.1	-	-	-	2.8		0.4	0.5	0.3
Gondoic acid (C20:1)	-	-	1.3	-	-	-	0.2	2.8	0.2	-	-	-
Gadoleic acid (C20:1)	0.2	0.3	-		-	-	-	-	-	0.3	0.3	0.2
Behenic acid (C22:0)	-	0.1	-	0.2	-	-	-	-	-	0.1	0.2	0.4
Erucic acid (C22:1)	-	-	-	-	-	-	-	0.4	-	-	0.2	-
∑PUFAs	10.1	6.2	29.5	16.5	24.8	29.0	36.4	12.0	75.8	56.7	52	61.9
∑MUFAs	72.6	79.4	64.0	74.6	67.2	60.6	55.8	76.5	14.7	29.8	29.8	23.9
∑SFAs	17.2	14.4	4.4	8.5	7.2	6.4	4.4	8.8	9.5	13.2	15.1	14.2

O: Oleic acid; L: Linoleic acid; P: Palmitic acid; PUFAs: polyunsaturated fatty acids; MUFAs: monounsaturated fatty acids; SFAs: saturated fatty acids. [a] Some irrelevant fatty acids are not presented in this table.

3.2. Experimental Dissolving Power of Refined Vegetable Oils and Mixtures with Their Derivatives

3.2.1. Extractability of Total Phenolic Compounds

The TPC in refined vegetable oils and mixtures with their derivatives as solvents quantified by the Folin–Ciocalteu test is illustrated in Figure 3. Generally, these refined oils consisting of more than ≥98% TAGs are often considered as strictly non-polar liquid solvents which are unfavorable for extraction. Therefore, only 1/3 of the refined oil solvents (avocado, olive, corn, and soybean) in this study showed relatively higher TPC yield, among which soybean oil obtained the highest (Figure 3a).

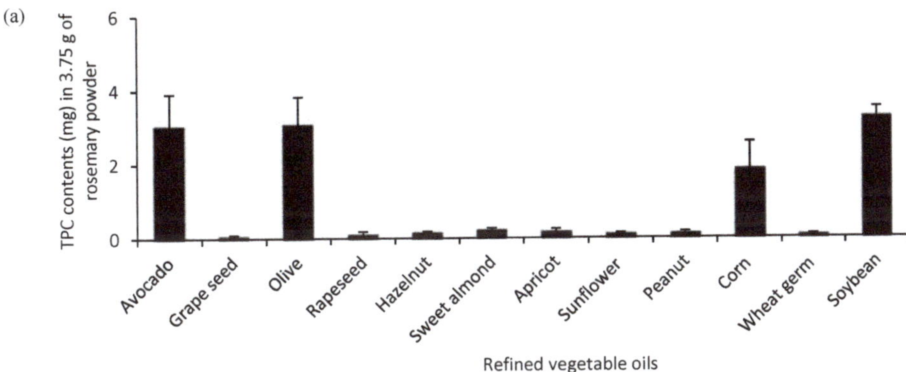

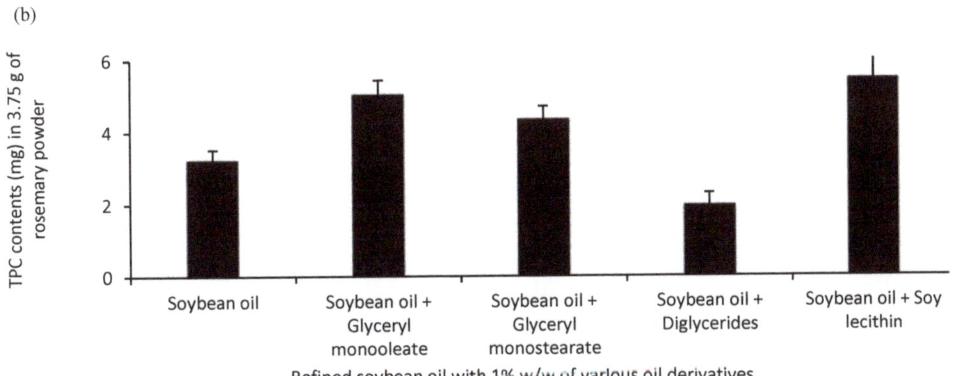

Figure 3. Total phenolic content (TPC) in rosemary oleo-extracts by using (**a**) refined vegetable oils and (**b**) refined soybean oil with the addition of oil derivatives as solvents.

This could be explained by its highest content of polyunsaturated fatty acids, allowing it to give itself a lower viscosity corresponding to a higher diffusivity that helps to increase the extraction yield. However, it was found that there is no obvious correlation between the TPC extraction yield (0.07–3.25 mg in 3.75 g of rosemary powders) and the polyunsaturation level of refined oil solvents (6.2–75.8%), which indicates that the TPC extraction yield may be more related to the presence of endogenous amphiphilic compounds like partial glycerides and phospholipids, and the refining degree of oils as well. For this, with the aim of further improving the dissolving power of refined soybean oils, the influence of oil derivatives on the TPC extraction was studied at the same dosage (1% *w/w*). Compared to the refined soybean oil, the addition of amphiphilic oil derivatives helped to increase the TPC extraction efficiency with the exception of DAG (Figure 3b). The performance of these food additives was found corresponding to the order of their hydrophilicity: soy lecithin > GMO > GMS > DAG. The presence of more polar phosphate groups in the soy lecithin allowed the formation of more reverse micelles containing phenolic compounds in the water pool center [26], which led to the highest TPC increment by approximately a factor of 1.7. Because of the higher lipophilicity of DAG, it obtained a lower TPC yield than MAG. GMO had a higher TPC extraction yield than that of GMS, which could be explained by the fact that the unsaturated GMO with a higher polarity and lower viscosity could facilitate the TPC extraction.

3.2.2. Extractability of Carnosol, Carnosic, and Rosmarinic Acids

Specifically, the content of main antioxidants in the refined oil solvents and their mixtures with oil derivatives is presented in Figure 4. All refined vegetable oils used were capable of extracting apolar carnosol and carnosic acid to various degrees, whereas polar rosmarinic acid was undetectable in all oil solvents. Similar to TPC results, the highest content of carnosic acid (1.76 mg in 100 g oil) and carnosol (0.16 mg in 100 g oil) was also observed in refined soybean oils (Figure 4a), which was thus selected for further extraction improvement with the addition of 1% *w/w* of oil derivatives. Compared to aqueous methanolic rosemary extracts, oily solvents could achieve a higher extraction yield of main antioxidants in total, although extracts by organic solvents obtained considerable carnosol and rosmarinic acids (Figure 4b).

(a)

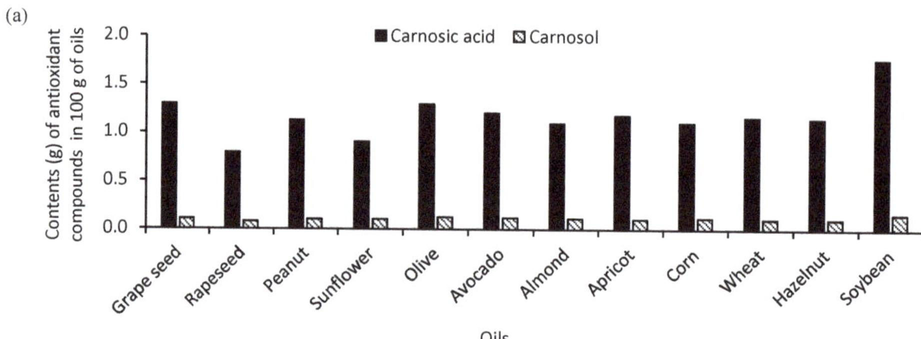

(b)

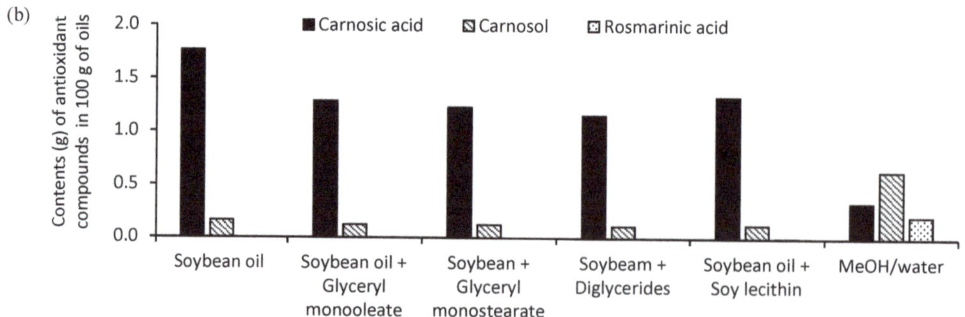

Figure 4. Content of carnosol and of carnosic and rosmarinic acids in rosemary oleo-extracts determined by HPLC using (**a**) refined vegetable oils and (**b**) refined soybean oil with the addition of oil derivatives as solvents.

Soy lecithin was the most suitable additive for the extraction of main rosemary antioxidants, followed by GMO, GMS, and DAG. Although the content of carnosic acid and carnosol in the solvent system was not as much as that in refined soybean oils, it is interesting to quantitatively detect rosmarinic acids within. Its higher extractability for both polar and non-polar compounds could also be the reason why the solvent system gave the highest TPC content compared to other oily solvents. Considering the relatively high extractability to phenolic compounds, refined soybean oil and its mixture with soy lecithin could be considered as good solvents for the extraction of rosemary antioxidant compounds.

3.2.3. Extractability of Volatile Aroma Compounds (VACs)

The standard of seven VACs at the same different concentration gradients was analyzed first in different aromatized oils under the same HS-SPME/GC-MS conditions in order to evaluate the extraction efficiency of the oily matrix and HS-SPME fiber. The fiber used was kept consistent and updated for analyses of each independent aromatized oil in order to minimize the error quantification. The content of major VACs in all oily solvents is presented in Figure S2, Supplementary Materials.

Major monoterpene hydrocarbons (e.g., α-pinene, limonene, and *o*-cymene) and highly odoriferous oxygenated monoterpenes (e.g., borneol, camphor, eucalyptol, and terpinen-4-ol) were both quantified in all oily extracts, but their contents were different from those in rosemary essential oils [27]. Eucalyptol was found as the major compound in oily solvents with the highest concentration, followed by camphor, α-pinene, *o*-cymene, and limonene. Different refined oil solvents showed an individual difference regarding the proportion of major VACs extracted (Figure S2a, Supplementary Materials). In other words, the highest contents of major VACs observed here were eucalyptol (49.26% *w/w*) in wheat germ oil, camphor (13.48% *w/w*) in rapeseed oil, α-pinene in soybean oil (10.70% *w/w*), terpinen-4-ol (3.65% *w/w*) in corn oil, as well as limonene (7.96% *w/w*), *o*-cymene (9.27% *w/w*), and borneol (3.05% *w/w*) in hazelnut oil. Furthermore, the mixture of refined soybean oil and 1% w/w of oil derivatives was studied as well to verify its performance consistency (Figure S2b, Supplementary Materials). Compared to refined soybean oil solely, the addition of oil derivatives could help to significantly increase the extraction yield of eucalyptol and α-pinene than other VACs. The highest content of camphor (16.61% *w/w*), terpinen-4-ol (4.24% *w/w*), and borneol (3.66% *w/w*) was found in the mixture of refined soybean oil and DAG. Therefore, it is worthy to note that adding oil derivatives to refined oil solvents could increase the extraction selectivity of VACs, which gave a higher yield of total oxygenated monoterpenes than total monoterpene hydrocarbons compared to using refined oils alone (Figure S2c, Supplementary Materials).

3.3. Theoretical Dissolving Power of Refined Soybean Oils and Mixtures with Their Derivatives

From experimental results above, it can be noted that oily solvents with various solutes are multi-component systems for which it is difficult to predict the dissolving power due to their complexities. Conventional solubility simulation methods (Hilderbrand or Hansen solubility parameters, etc.) often underestimate molecular structural difference and interaction force (type, proportion, and position of three main fatty acid chains in TAG, etc.) resulting in predicted results with little differences between each other. Thus, a powerful COSMO-RS simulation was conducted to determine the relative solubility of both phenolic antioxidants and VACs in refined soybean oils and their mixtures with oil derivatives. The σ-potentials of solvents and solutes (i.e., TAGs, oil derivatives, major antioxidants, and VACs) derived from the COSMO-RS simulation can be successfully employed to describe and classify solvents in a purely predictive manner with a good consideration of hydrogen bond donor–acceptor interactions. Generally, the region σ ± 0.01 e/A^2 is considered as non-polar or weakly polar. For instance, the σ-profile of limonene showed two peaks resulting from the hydrogen atoms on the negative side and from the carbon atoms on the positive side. Therefore, the σ-potential of limonene was similar to the U-shape centered at σ = 0, which is the typical characteristic of non-polar solvents. As represented in Table 2, the TAG composition possibilities in refined soybean oils including 66% of TAG 1 (R1: C18:2, R2: C18:2, R3: C18:2), 23% of TAG 2 (R1: C18:1, R2: C18:2, R3: C18:2), and 11% of TAG 3 (R1: C18:2, R2: C18:1, R3: C16:0), and their mixtures with oil derivatives could be taken into consideration in this quantic chemical approach. As the logarithm of the best solubility is set to 0, all these solvents could be predicted under simulated industrial conditions at 40 °C. It was found that the relative solubility of all components was below zero, indicating that these oily solvents may not be perfect solvents theoretically though they have considerable dissolving power in real extractions. Compared to the relative solubility value of refined soybean oil as the reference solvent, oily solvents with higher relative solubility values (in green) could be recognized as having better dissolving powers. Otherwise, solvents with poorer dissolving capacities had lower relative solubility values (in red). The theoretical solubility of both non-volatile antioxidants and VACs in the mixture of refined soybean oil and oil derivatives was in good consistency with experimental data, with the exception of camphor in refined soybean oils with DAG. Overall, refined soybean oils with soy lecithin (1%, *w/w*) could be theoretically considered as the optimal solvent because of its near-zero relative solubility for all bioactive compounds, which is also in good accordance with experimental results obtained previously.

Table 2. Relative solubility of non-volatile and volatile bioactive compounds in refined soybean oil and its mixture with oil derivatives using the COSMO-RS approach.

Bioactive Compound	Oily Solvents				
	Refined Soybean Oil (RSO) [a]	Addition of Oil Derivatives (1%, *w/w*)			
		RSO + DAG	RSO + GMO	RSO + GMS	RSO + Soy lecithin
Carnosic acid	−2.7222	−2.7173	−2.7066	−2.7101	−2.3683
Rosmarinic acid	−2.5669	−2.5584	−2.5396	−2.5480	−1.9897
Carnosol	−1,9426	−1.9408	−1.9358	−1.9370	−1.8102
Borneol	−1.5602	−1.5289	−1.5257	−1.5259	−1.4458
Camphor	−0.7953	−0.7956	−0.7946	−0.7932	−0.7931
o-Cymene	−0.2403	−0.2407	−0.2411	−0.2413	−0.2409
Eucalyptol	−1.1891	−1.1889	−1.1869	−1.1819	−1.1885
Limonene	−0.3473	−0.3477	−0.3483	−0.3485	−0.3484
α-pinene	−0.6353	−0.6358	−0.6368	−0.6369	−0.6378
Terpinen-4-ol	−1.3907	−1.3896	−1.3859	−1.3827	−1.3395

[a] Refined soybean oil: 66% TAG 1 (R1 = C18:2, R2 = C18:2, R3 = C18:2), 23% TAG 2 (R1 = C18:1, R2 = C18:2, R3 = C18:2), 11% TAG 3 (R1 = C18:2, R2 = C18:1, R3 = C16:0); DAG: diglycerides; GMO: glyceryl monooleate; GMS: glyceryl monostearate. Grey: Reference; Green: Better or equivalent than reference; Red: Worse than reference.

3.4. Classification of Oily Solvents

Generally, the variability of independent principal components is the percentage of information which is well represented for other components. In this case, the PCA plotted all oily solvents with the main components PC1, PC2, and PC3 representing 95.27% of the original information with less than 5% loss of information (Figure 5). The distribution of 16 oily solvents on the 2D mapping graphs (PC1 versus PC2 and PC1 versus PC3) described well their inner similarity depending on the experimental solubility of TPC, non-volatile antioxidants, and VACs inside, which is directly related to independent variables that can bring significant information for the oily solvent discrimination. The following clustering analysis was carried out on the basis of the concentration of both non-volatile and volatile bioactive compounds as well (Figure 6). The dendrogram showed that the dissolving power of oily solvents could be finally aggregated into two main clusters. The first cluster including all refined vegetable oils showed the highest dissimilarity as compared to the second cluster, which indicated the significant effect of the addition of oil derivatives to refined oil solvents on their extraction improvement.

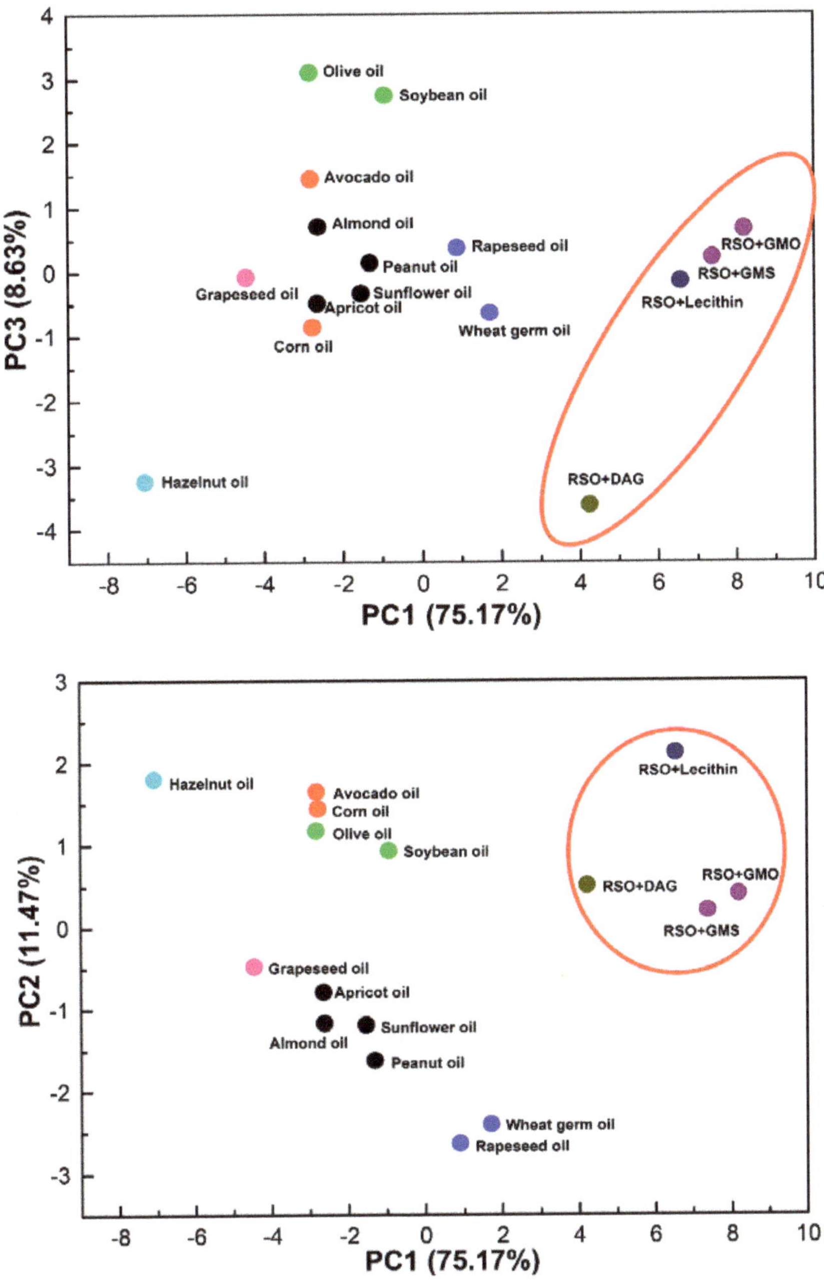

Figure 5. Dissolving power of oily solvents in principal component analytical plots (PC1 *versus* PC2 and PC1 *versus* PC3) corresponding to volatile and non-volatile bioactive compounds extracted.

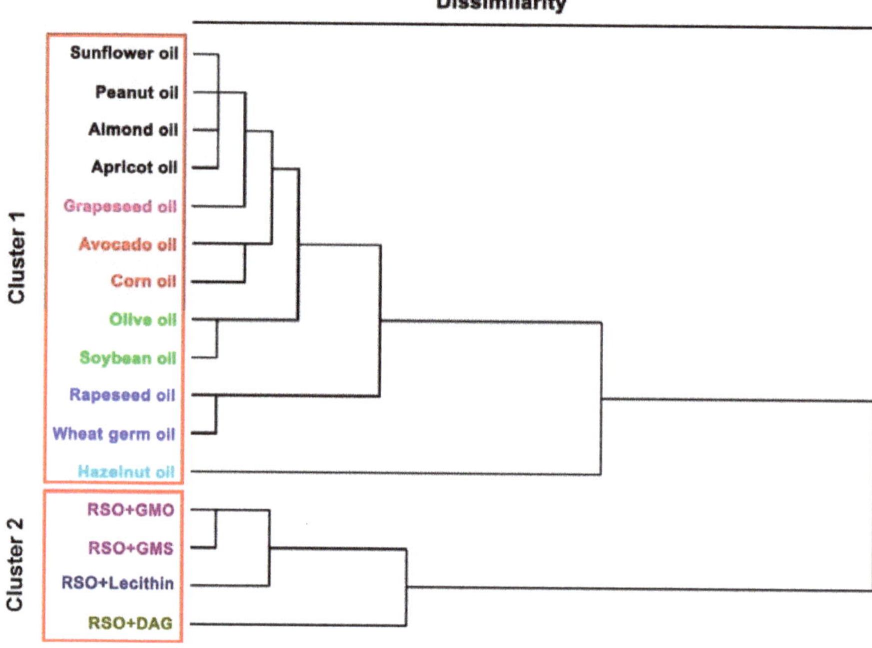

Figure 6. Dendrogram resulting from Ward's hierarchical cluster analysis corresponding to the dissolving power of oily solvents in terms of volatile and non-volatile bioactive compounds extracted.

The oily solvents merging together within each cluster showed low dissimilarity on the dissolving power. Therefore, the low dissimilarity was found in cluster 1 among several sub-clusters in the same color (refined sunflower, peanut, almond, and apricot oils, etc.), and in cluster 2 between refined soybean oil with GMO and GMS. However, some oily solvents presented high dissimilarity corresponding to their lower or higher dissolving powers as compared to others. For instance, refined wheat germ and rapeseed oils in the same sub-cluster had nearly the same dissolving power with much lower yields for all bioactive compounds extracted, although they had a relatively higher affinity to VACs. Refined hazelnut oil with the lowest dissolving power showed the highest dissimilarity, which was isolated at a distance from others in the PCA graphs. A similar case happened in another cluster for refined soybean oil with DAG, which also had the lowest dissolving power. Nonetheless, refined soybean oil with soy lecithin showed a high dissimilarity because of its best performance in the oleo-extraction of both non-volatile and volatile bioactive compounds from rosemary. Interestingly, refined vegetable oils having similar TAG composition were distributed in different sub-clusters, which also further states the fact that the dissolving power of these complex solvents is more correlated to their endogenous amphiphilic constitutes. Refined sunflower oil was previously reported as the optimal product for the extraction of six VACs from basil [24]. Nevertheless, refined soybean oil had a better dissolving power in this work for the maximal extraction of non-volatile antioxidants and VACs from rosemary, indicating the specificity of oleo-extraction that is worthy for case-by-case studies. All results were in good consistency with the PCA observation and theoretical relative solubility results, which further proved the effectiveness of COSMO-RS for a priori solvent screening before trial and error.

Since the dissolving power of vegetable oils have been confirmed as alternative solvents for natural product extractions, a novel oleo-extraction process could be subsequently developed based on green extraction principles with the integration of innovative techniques (ultrasound, microwave, etc.), resulting in a high-efficiency, time- and energy-saving extraction, as well as high value-added lipidic products with the maximal risk reduction of oxidation or degradation thanks to the extracted bioactive compounds within [19,28]. In addition, vegetable oils could also be extracted by other alternative eco-friendly solvents (terpenes, 2-MeTHF, liquefied gas, etc.) to n-hexane, which conceivably led to a truly green extraction process for future industrial applications [29–32].

4. Conclusions

This study aimed at developing a green oleo-extraction of total volatile and non-volatile bioactive compounds from rosemary leaves towards a zero-waste biorefinery concept using vegetable oils and their amphiphilic derivative constitutes as bio-based solvents. The experimental results showed that refined soybean oil performed the best concerning the yield of both major phenolic antioxidants and VACs in rosemary. Moreover, the addition of oil derivatives, soy lecithin in particular, surprisingly improved the extraction efficiency of more polar compounds and VACs on the basis of the original oleo-extraction. Meanwhile, a good consistency with COSMO-RS simulation proved its suitability for modeling these supramolecular solvent systems, which could overcome the limit of conventional solubility methods. Considering theoretical, experimental, and statistical results, refined soybean oil with soy lecithin (1%, w/w) appears to be the most promising and economically viable solvent among all oily solvents tested for the maximal extraction of major antioxidants and VACs from rosemary.

From the practical point of view, COSMO-RS could be conducted as a heuristic tool before real extractions for a low-cost and high-accuracy screening of complex solvents or multicomponent solute–solvent mixtures. As the dissolving power of vegetable oils as bio-based solvents has been gradually proved for solutes with a broader range of polarity, the green oleo-extraction updated in this study is more comprehensive, where no additional separation steps are required after extractions. Revisiting this safe, easy-to-use, and eco-friendly method has produced novel enriched oils, thereby providing a relatively green solution towards the biorefinery of wastes from the processing of nearly all plant materials, depending on the market demand in the cosmetic and agro-food industries.

Supplementary Materials: The following are available online at http://www.mdpi.com/2076-3921/8/5/140/s1, Figure S1: Three-dimensional chemical structures and σ–surfaces of both volatile and non-volatile solutes, triglyceride possibilities in refined soybean oils and oil amphiphilic derivatives generated by COSMO-RS, Figure S2: Major volatile aroma compounds (VACs) in rosemary oleo-extracts determined by HS-SPME/GC-MS using (a) refined vegetable oils and (b) refined soybean oil with addition of oil derivatives as solvents, (c) the content of total monoterpenes *versus* total oxygenated monoterpenes in various oily solvent systems.

Author Contributions: Y.L., A.-S.F.-T. and F.C. designed the project, Y.L., K.B., N.R. and A.-S.F.-T. conducted the experiments and data analysis, R.B. and Y.W. provided technical support, Y.L., K.B., A.-S.F.-T., F.C. wrote the paper.

Funding: This research was funded by the National Natural Science Foundation of China with grant number [31701633] and the Program for Guangdong YangFan Introducing Innovative and Entrepreneurial Teams with grant number [2016YT03H132].

Acknowledgments: The authors would like to thank Jinan University (Guangzhou, China), Avignon University (France), INRA Avignon (France), Prince of Songkla University (Thailand), and Naturex (France), for the contribution to this study.

Conflicts of Interest: The authors declare no conflict of interest.

References

1. Pfaltzgraff, L.A.; De bruyn, M.; Cooper, E.C.; Budarin, V.; Clark, J.H. Food waste biomass: A resource for high-value chemicals. *Green Chem.* **2013**, *15*, 307–314. [CrossRef]

2. Lin, C.S.K.; Pfaltzgraff, L.A.; Herrero-Davila, L.; Mubofu, E.B.; Abderrahim, S.; Clark, J.H.; Koutinas, A.A.; Kopsahelis, N.; Stamatelatou, K.; Dickson, F.; et al. Food waste as a valuable resource for the production of chemicals, materials and fuels. Current situation and global perspective. *Energy Environ. Sci.* **2013**, *6*, 426–464.

3. Chandel, A.K.; Garlapati, V.K.; Singh, A.K.; Antunes, F.A.F.; Silvério da Silva, S. The path forward for lignocellulose biorefineries: Bottlenecks, solutions, and perspective on commercialization. *Bioresour. Technol.* **2018**, *264*, 370–381. [CrossRef]

4. Kiss, A.A.; Lange, J.P.; Schuur, B.; Brilman, D.W.F.; van der Ham, A.G.J.; Kersten, S.R.A. Separation technology-making a difference in biorefineries. *Biomass Bioenergy* **2016**, *95*, 296–309. [CrossRef]

5. Procopio, A.; Alcaro, S.; Nardi, M.; Oliverio, M.; Ortuso, F.; Sacchetta, P.; Pieragostino, D.; Sindona, G. Synthesis, Biological Evaluation, and Molecular Modeling of Oleuropein and Its Semisynthetic Derivatives as Cyclooxygenase Inhibitors. *J. Agric. Food Chem.* **2009**, *57*, 11161–11167. [CrossRef] [PubMed]

6. Rapinel, V.; Rombaut, N.; Rakotomanomana, N.; Vallageas, A.; Cravotto, G.; Chemat, F. An original approach for lipophilic natural products extraction: Use of liquefied *n*-butane as alternative solvent to *n*-hexane. *LWT Food Sci. Technol.* **2017**, *85 Pt B*, 524–533. [CrossRef]

7. Chemat, F.; Abert-Vian, M.; Cravotto, G. Green extraction of natural products: Concept and principles. *Int. J. Mol. Sci.* **2012**, *13*, 8615–8627. [CrossRef] [PubMed]

8. Achat, S.; Toamo, V.; Madani, K.; Chibane, M.; Elmaataoui, M.; Dangles, O.; Chemat, F. Direct enrichment of olive oil in oleuropein by ultrasound-assisted maceration at laboratory and pilot plant scale. *Ultrason. Sonochem.* **2012**, *19*, 777–786. [CrossRef] [PubMed]

9. Herrero, M.; Ibañez, E. Green extraction process, biorefineries and sustainability: Recovery of high added-value products from natural sources. *J. Supercrit. Fluid* **2018**, *134*, 252–259. [CrossRef]

10. Allaf, T.; Tomao, V.; Ruiz, K.; Bachari, K.; ElMaataoui, M.; Chemat, F. Deodorization by instant controlled pressure drop autovaporization of rosemary leaves prior to solvent extraction of antioxidants. *LWT-Food Sci. Technol.* **2013**, *51*, 111–119. [CrossRef]

11. Jacotet-Navarro, M.; Rombaut, N.; Fabinao-Tixier, A.S.; Danguien, M.; Billy, A.; Chemat, F. Ultrasound versus microwave as green processes for extraction of rosmarinic, carnosic and ursolic acids from rosemary. *Ultrason. Sonochem.* **2015**, *27*, 102–109. [CrossRef] [PubMed]

12. Tommasi, E.; Cravotto, G.; Gallettim, P.; Grillo, G.; Mazzotti, M.; Sacchetti, G.; Samorì, C.; Tabasso, S.; Tacchini, M.; Tagliavini, E. Enhanced and Selective Lipid Extraction from the Microalga P. tricornutum by Dimethyl Carbonate and Supercritical CO2 Using Deep Eutectic Solvents and Microwaves as Pretreatment. *ACS Sustain. Chem. Eng.* **2017**, *5*, 8316–8322. [CrossRef]

13. Brewer, M.S. Natural antioxidants: Sources, compounds, mechanisms of action, and potential applications. *Compr. Rev. Food Sci. F* **2011**, *10*, 221–247. [CrossRef]

14. Erkan, N.; Ayranci, G.; Ayranci, E. Antioxidant activities of rosemary (*Rosmarinus Offcinalis* L. extract), blackseed (*Nigella sativa* L.) essential oil, carnosic acid, rosmarinic acid and sesamol. *Food Chem.* **2008**, *110*, 76–82. [CrossRef] [PubMed]

15. Frankel, E.N.; Huang, S.W.; Aeschbach, R.; Prior, E. Antioxidant activity of a rosemary extract and its constituents, carnosic acid, carnosol, and rosmarinic acid, in bulk oil and oil-in-water emulsion. *J. Agric. Food Chem.* **1996**, *44*, 131–135. [CrossRef]

16. Moreno, S.; Scheyer, T.; Romano, C.S.; Vojnov, A.A. Antioxidant and antimicrobial activities of rosemary extracts linked to their polyphenol composition. *Free Radic. Res.* **2006**, *40*, 223–231. [CrossRef] [PubMed]

17. Jacotet-Navarro, M.; Laguerre, M.; Fabiano-Tixier, A.S.; Tenon, M.; Feuillère, N.; Billy, A.; Chemat, F. What is the best ethanol-water ratio for the extraction of antioxidants from rosemary? Impact of the solvent on yield, composition, and activity of extracts. *Electrophoresis* **2018**, *39*, 1946–1956. [CrossRef] [PubMed]

18. Boutekedjiret, C.; Bentahar, F.; Belabbes, R.; Bessiere, J.M. Extraction of rosemary essential oil by steam distillation and hydrodistillation. *Flavour Fragr. J.* **2003**, *18*, 481–484. [CrossRef]

19. Yara-Varón, E.; Li, Y.; Balcells, M.; Canela-Garayoa, R.; Fabinao-Tixier, A.S.; Chemat, F. Vegetable oils as alternative solvents for green oleo-extraction, purification and formulation of food and natural products. *Molecules* **2017**, *22*, 1474. [CrossRef] [PubMed]

20. Figoli, A.; Marino, T.; Simone, S.; Di Nicolo, E.; Li, X.M.; He, T.; Tornaghi, S.; Drioli, E. Towards non-toxic solvents for membrane preparation: A review. *Green Chem.* **2014**, *16*, 4034–4059. [CrossRef]

21. Moity, L.; Molinier, V.; Benazzouz, A.; Joossen, B.; Gerbaud, V.; Aubry, J.M. A "top-down" in silico approach for designing ad hoc bio-based solvents: Application to glycerol-derived solvents of nitrocellulose. *Green Chem.* **2016**, *18*, 3239–3249. [CrossRef]

22. Byrne, F.P.; Jin, S.; Paggiola, G.; Petchey, T.H.M.; Clark, J.H.; Farmer, T.J.; Hunt, A.J.; McElroy, C.R.; Sherwood, J. Tools and techniques for solvent selection: Green solvent selection guides. *Sustain. Chem. Process.* **2016**, *4*, 7–31. [CrossRef]

23. Li, Y.; Fabiano-Tixier, A.S.; Ruiz, K.; Castera, A.R.; Bauduin, P.; Diat, O.; Chemat, F. Comprehension of direct extraction of hydrophilic antioxidants using vegetable oils by polar paradox theory and small angle X-ray scattering analysis. *Food Chem.* **2015**, *173*, 873–880. [CrossRef]

24. Pingret, D.; Durand, G.; Fabiano-Tixier, A.S.; Rockenbauer, A.; Ginies, C.; Chemat, F. Degradation of edible oil during food processing by ultrasound: Electron paramagnetic resonance, physicochemical, and sensory appreciation. *J. Agric. Food Chem.* **2012**, *60*, 7761–7768. [CrossRef] [PubMed]

25. Xenakis, A.; Papadimitriou, V.; Sotiroudis, T.G. Colloidal structures in natural oils. *Curr. Opin. Colloid Interface Sci.* **2010**, *15*, 55–60. [CrossRef]

26. Li, Y.; Fabiano-Tixier, A.S.; Chemat, F. Vegetable oils as alternative solvents for green extraction of natural products. In *Edible Oils: Extraction, Processing and Applications*; Chemat, S., Ed.; CRC Press, Taylor & Francis: Boca Raton, FL, USA, 2017; pp. 205–222.

27. Filly, A.; Fernandez, X.; Minuti, M.; Visinoni, F.; Cravotto, G.; Chemat, F. Solvent-free microwave extraction of essential oil from aromatic herbs: From laboratory to pilot and industrial scale. *Food Chem.* **2014**, *150*, 193–198. [CrossRef] [PubMed]

28. Goula, A.M.; Ververi, M.; Adamopoulou, A.; Kaderides, K. Green ultrasound-assisted extraction of carotenoids from pomegranate wastes using vegetable oils. *Ultrason. Sonochem.* **2017**, *34*, 821–830. [CrossRef]

29. Sicaire, A.G.; Abert-Vian, M.; Filly, A.; Li, Y.; Billy, A.; Chemat, F. 2-Methyltetrahydrofuran: Main properties, production processes, and application in extraction of natural products. In *Alternative Solvents for Natural Products Extraction*; Chemat, F., Albert-Vian, M., Eds.; Springer: Berlin/Heidelberg, Germany, 2014; pp. 253–268.

30. Li, Y.; Fabiano-Tixier, A.S.; Ginies, C.; Chemat, F. Direct green extraction of volatile aroma compounds using vegetable oils as solvents: Theoretical and experimental solubility study. *LWT Food Sci. Technol.* **2014**, *59*, 724–731. [CrossRef]

31. Li, Y.; Fine, F.; Fabiano-Tixier, A.S.; Abert-Vian, M.; Carre, P.; Pages, X.; Chemat, F. Evaluation of alternative solvents for improvement of oil extraction from rapeseeds. *C. R. Chim.* **2014**, *17*, 242–251. [CrossRef]

32. Rapinel, V.; Santerre, C.; Hanaei, F.; Belay, J.; Vallet, N.; Rakotomanomana, N.; Vallageas, A.; Chemat, F. Potentialities of using liquefied gases as alternative solvents to substitute hexane for the extraction of aromas from fresh and dry natural products. *C. R. Chim.* **2018**, *21*, 590–605. [CrossRef]

MDPI
St. Alban-Anlage 66
4052 Basel
Switzerland
Tel. +41 61 683 77 34
Fax +41 61 302 89 18
www.mdpi.com

Antioxidants Editorial Office
E-mail: antioxidants@mdpi.com
www.mdpi.com/journal/antioxidants